Best Practices and Strategies for Career and Technical Education and Training

Best Practices and Strategies for Career and Technical Education and Training

A Reference Guide for New Instructors

KINGA N. JACOBSON

authorHOUSE®

AuthorHouse™
1663 Liberty Drive
Bloomington, IN 47403
www.authorhouse.com
Phone: 1-800-839-8640

Published by AuthorHouse   01/14/2013

ISBN: 978-1-4817-0637-7 (sc)
ISBN: 978-1-4817-0638-4 (e)

Library of Congress Control Number: 2013900551

This book is dedicated to my Professors, with respect, for igniting my passion for career and technical education and training and to my Family, with love, for supporting me in accomplishing my dreams.

CONTENTS

ABSTRACT

Best Practices and Strategies for Career and Technical Education and Training is a reference guide for novice instructors. The first chapter creates the context for the outlined instructional practices and strategies by providing a basic overview of the mission and goals of career and technical education and the evolution of training and development. The context is completed by a review of basic personality traits that affect the instructional style of the facilitator. Chapter two revolves around best practices instructors use in classroom management and assessment and includes an extensive analysis of the course syllabi and calendar which resemble procedures applied by managers in effective business operations. Chapter three explores basic career and technical education and corporate training instructional tools. It provides a practical reference of student engagement strategies and leadership techniques applicable in the career and technical education and corporate training setting.

CHAPTER 1

The Context of Career and Technical Education and Training

The objectives of this chapter are to provide instructors a basic understanding of:

- the mission and goals of career and technical education
- the evolution and role of corporate training and development
- the personality traits that influence educator and trainer instructional practices

1.1. The Mission and Goals of Career and Technical Education

Career and technical education, an essential part of the American education system, prepares students for the world of work. It is defined as education that prepares individuals to fulfill their working potential by providing the practical experience needed for employment success (Scott, 2008).

The mission of career and technical education is to "prepare all individuals for lasting and rewarding employment, further education and lifelong learning" (Gordon, 2008).

Today's knowledge and service based economy requires that everyone has adequate technical skills. Career and technical education plays an important role in making sure that individuals, companies and the national economy remain competitive in this context. The development of skills and knowledge for gainful employment and ongoing, self-directed learning are keys in this education. Career and technical education is essential for "addressing career needs to upgrade, retain and maintain occupational choice and technological literacy for all members of society" (Scarcella, 2012). Having a proactive mentality and a flexible and practical approach, career and technical education provides what learners need for career success and plays a foundational role in the overall betterment of society and economy.

Historically, career and technical education rooted in vocational training for development of entry level career skills, was fairly narrowly focused on specific technology job skills (Woodward, 1980). Today, the programs take a broader approach that includes academic preparation as well because basic academic skills like mathematics and English and personal qualities such as critical thinking, teamwork and professionalism are considered essential for success in the current global environment. Thus, skills such as active listening, conflict resolution or interpersonal communications are more important than ever. The mission evolved, the emphasis moving toward an inclusive, lifelong educational approach which prepares for all aspects of a career rather than just entry-level employment. As individuals are expected to change jobs and careers several times during their lifespan, the basic competencies, once the exclusive focus of career and technical education, are now complemented by core abilities essential for success in any industry. In flexible, flat organizational structures where teamwork, problem solving and decision making are essential, core abilities and strong academics become, as Barlow points out, a national concern (Scott, 2008). Consequentially, career and technical education program content now includes these topics, focusing on what is truly valued by business and society to achieve satisfaction of genuine economic needs. Current workplace demands go far beyond the ability to perform a certain well-defined job, new technologies and the fast pace of change requiring us to constantly evaluate and adjust. To prepare individuals for career success, career and technical education needs to create good communicators and readily employable thinkers who embrace change and permanent development.

In this context and in alignment with its ultimate goal and mission career and technical education must remain proactive in reducing and preventing skilled worker shortages and preparing graduates who, via gainful employment, accelerate the nation's economic vitality. Career and technical education, as a vital link between business, communities and education systems as well as between secondary and four-year postsecondary education, needs to stay focused on developing, in addition to technical and academic skills necessary for career success, learners' characters, facilitating their growth in human relations and social citizenship.

Career and technical instruction is practical, applicable in the real world of work and inclusive of all learners. Instruction is delivered in a student focused learning community in which all backgrounds and experience levels find their role and come together to build fruitful collective intelligence. Specific instructional and assessment practices lead to student responsibility and accountability while accommodating various ability levels, cultural and economic backgrounds and individual goals. Career and technical education sees teaching and learning as an interconnected process in which the subject matter expertise is only the beginning, seeking to engage students through rich teaching methods incorporating modern technology, relevant academics and practical scenarios. Analytical and problem solving exercises construct learner knowledge and refine thinking and lead to amplified employability. Career and technical education aims to reach broad populations and to provide quality education accessible to all, having a "concern for the quality of human existence" and keeping in mind learners' socio-economic demographics and affordability needs (Scott, 2008). Its strategy is to employ distance delivery methods, give credit for prior work based learning and provide opportunities for acquiring skills via apprenticeships and mentorships. Programs reinforce classroom education in laboratory and worksite settings. The experiences go beyond simply experimenting with tasks and machinery, to asking learners to engage with their learning at a deep, evaluative level so that they can develop long term, easily transferable knowledge.

Career and technical education programs offer a comprehensive curriculum that includes technical knowledge, employability skills and academics foundational for success in the real world of work. The curriculum includes a sequence of general courses referred to as "common core" as well as program specific occupational courses. In this, it shares responsibility with other educators for exceptional curriculum quality that brings learner readiness for higher educational levels and lifelong skills and knowledge improvement. By means of a challenging, engaging, effective and relevant curriculum students are ready to fill available job openings and further their education, the ultimate goal being to provide opportunities for discovery of individual potential, creativity and talent for continuous development in a technological and global society. This means that the programs are in close correlation with the needs of the economy while also serving individual career goals. Career and technical education is an authentic learning climate that assures economic vitality and competitiveness of the economic environment (Scott, 2008).

Career and technical education programs are delivered in high school and two-year post-secondary settings and typically include dual credit options, certificates and associate degrees. In the K-12 environment the courses are integrated with the comprehensive high school curriculum while at the post-secondary level they are accessible via two-year community and technical colleges and adult education centers. Their goal is to help students understand and prepare for the occupational options available to them by experiencing various career related job tasks.

The educational setting is a mix between theory and practice often involving off site work experiences such as clinics or shops. Instructors as well as mentors are selected carefully for their industry related expertise, field experience and willingness to create a collaborative, supportive learning environment. Classes are usually small in size and delivery methods range from face-to-face, blended, video-conference to fully online or self-paced. In general, great attention is given to flexibility and accommodation of diverse learner needs while maintaining the quality of the educational outcome.

Career and technical education's target audience is the population seeking employment. The career and technical education program population ranges from high school students with limited work-life experience to retired adults looking to refresh their skills. Within this continuum, there are learners who already hold part or full time entry or mid-level jobs and there are experienced, specialized professionals looking for continuing education and to upgrade their skills and knowledge. This very diverse student audience includes adults who juggle multiple family and life responsibilities including work and home. Learners also often have some level of practical job experience or even a completed educational degree. Some struggle with financial burdens, job loss, family issues and other distracters affecting their ability to succeed academically. Also, many career and technical education students come from diverse and first generation college education families that do not support degree completion. Due to these circumstances, learning must happen in context of everything else they need to accomplish and expand themselves to with their limited resources. Even though most of them enter programs with the mindset of earning an associate degree or certificate, many leave without completing a credential. In fact, nearly three fourth of career and technical education students complete less than eight months of post-secondary coursework. Bailey et al. (2004) present information on the characteristics of occupational students enrolled in community college programs. The areas of study with the largest enrollments include business and office, health and computer technologies. The study shows that in occupational programs there are fewer females and the majority of learners are from economically disadvantaged minority populations. Learners tend to be older than 24 years of age and are less likely to attend school full-time (Bailey et al., 2004). In this context, it is natural that career and technical education evolved to be highly supportive and learner centered, priority being given to using collaborative strategies that increase student success.

Students are of different ages and gender, ethnic and religious background and various levels of experience. In a multicultural society career and technical education incorporates the values and needs of diverse learners and builds career pathways that have long lasting impact on society, its role extending beyond the classroom to provide support and motivation for individuals in process of discovering their existential moment. It aligns its expectations with the students', acting consistently in support of their goals and advocates employment of women, minorities and older adults to promote workforce diversity. To remain true to its mission, career and technical education assesses performance via graduate and industry feedback and leads change in societal attitudes. Student by student and course by course, it constructs a better society, a stronger next generation and a more competitive future economy.

Ultimately, career and technical education helps students discover their individual strengths, interests and potential and allows them to fully engage in and own the achievement of their goals. The learners, entering with varying levels of ability and preparation have diverse needs and individual goals. However, career and technical education's comprehensive curriculum that develops subject matter expertise and includes supervised work experiences, lets students apply the skills they learn and realize their value and benefits. It is a setting that represents an "opportunity for meaningful learning, establishing or strengthening social connections, and preparing for the workplace or continuing education" (Scott, 2008).

The NAVE report by Silverberg et al. (2002) supports the view that career and technical education under current federal policy attempts to achieve multiple goals and objectives. In addition to the development of student technical and academic skills, career and technical education is expected to contribute to high school completion, entry to post-secondary education and training, post-secondary degree completion, and employment. To achieve this goal, it sets high expectations, is comprehensive, practical and work-centered, and focuses on learner collaboration and support.

In essence, career and technical education leads the development of future industry leaders; the goal of graduating career bound learners replacing the traditional entry level worker preparation. The programs of study and career pathways allow students to work on a continuum of skills representing a certain career path. Learners understand the value and application of their education in real life, allowing them to connect course content with their future societal roles (Gordon, 2008). Additionally, this instructional setting is focused on providing guidance and advice to learners with diverse needs, including services such as special needs accommodations, tutoring and mental health counseling.

Career and technical education aims creation of multi-skilled workers ready for winning the global competition and has been shown to increase student motivation and degree completion. Several recent studies found that career and technical education has a positive impact on high school graduation rates, labor market outcomes, and post-secondary enrollment (Scott, 2008). For instructors working in this setting, it is beneficial to have a philosophy to provide direction, values and benchmarks as they seek to lead students to long term career success. Having a personal philosophy helps clarify what they need to teach, to whom, how and why and leads to specific classroom management and curriculum development approaches. Philosophy is defined as "a system of principles for guidance in practical affairs" (Merriam-Webster, 2012). The educator's philosophy relates to their own personal system and conceptions, beliefs, and values in relation to perceiving the world and themselves as professionals. Career and technical educator philosophical assumptions revolve around developing individuals who contribute to society and also maximize their individual talents and interests. Calhoun and Finch (1982) show that there needs to be cooperation, rather than dualism, between career and technical educational approaches, needing to have a constant focus on satisfying the community's job market needs, social utility and the betterment of the national economy (Scott, 2008).

Lynch (2000) indicates that a new vision is needed in career and technical education. Lynch pictures the new career and technical education as being academically rigorous and career relevant, teaching all aspects of the employment and continuing education. The author identifies the purposes of career and technical education as providing career exploration and planning and enhancing academic achievement and motivation via generic work competencies and lifelong learning pathways for continuing education. The work states that career and technical education is "education through occupation serving as an instructional modality for traditional academic content" showing the utmost importance of close collaboration with industries so that career and technical programs produce graduates that are readily employable in today's and tomorrow's industries. This goal, the author explains, is accomplished through "rigorous, coherent, sequenced programs of study that include high-level academics, a body of knowledge recognized by industry standards, technology applications, employability skills and work-based learning in all aspects of the industry" (Lynch, 2000).

In conclusion, career and technical educators need to be aware of the socio-economic role they play and remain cognizant of the purpose of career and technical schooling. As Miller and Greggson (1999) indicate, instructors need to facilitate the "growth of learners who are competent problem solvers, makers of meaning, communicators, lifelong learners, citizens capable of adjusting to and promoting change, and practitioners of democratic processes proportionate for both academic and career and technical education" (Scott, 2008). Career and technical educators need to remain focused on developing, in addition to technical and academic skills, social citizenship traits necessary for career success and must embrace their roles in achievement of student dreams and the overall betterment of society and economy.

1.2. The Evolution and Role of Corporate Training and Development

Training and development refers to programs that aim the successful acclimation of new employees to the workplace and the assurance of continuously higher performance. Due to the constantly evolving, globally competitive marketplace, organizations are dependent on their highly skilled workforce to adapt, improve and stay efficient amidst controversial and changing expectations. As a result, these functions emerged to be recognized as integral parts of the corporate strategy. Most corporations use a comprehensive training and development approach to bring about the overall acclimation, improvement and education of their employees (Reference for Business, 2012). Training usually focuses on very specific, measurable goals such as learning to operate certain equipment or follow certain procedures. Development programs, on the other hand, concentrate on skills and abilities applicable across industries and careers, such as critical thinking, goal setting, decision making, or teamwork. In today's workplace, we frequently encounter new employee orientation, on-the-job training and skill improvement in teamwork, communications and leadership.

Although training and development is closely related to career and technical education instruction in general, the two programs most directly connected to the origins of vocational training are the employee acclimation and the on-the-job trainings. These evolve from the historical training structure called apprenticeship, in which the apprentice became part of the working team and transformed, over time, into a highly skilled master craftsman. As the society evolved into the Industrial Age of automated plants and large-scale production, the traditional apprenticeship close to disappeared, transforming into job task oriented vocational training. This type of development is well fitted for industrial needs geared toward division of labor and assembly lines needing highly productive workers with precise technical skills. However, automated production lines reduce worker satisfaction and skill transferability. Globalization induced a shift in jobs and industry needs, adaptability becoming key for long term competitiveness. In response, training and development reinvented itself to quickly and effectively respond to leadership expectations. Consumer demands, technology and the employee base to be trained became more diverse and more intense. The profession responded by launching training programs such as simulations, role plays and leadership development focusing on career coaching and adaptability to change in addition to the traditional orientation and on-the-job training (Reference for Business, 2012).

New employee orientation is concerned with introducing new hires to the company culture, organizational structure, leadership and general policies and procedures while on-the-job training focuses on development of skills and knowledge employees need to accomplish their position-related tasks. Typically, new employee orientation is delivered in a structured training room setting being exclusively a training and development function, while on the job training is held at least in part at the future work-site of the employee in collaboration with experienced mentors. The mentors, who are selected experts and long-term employees, guide the new team member through their growth and development process, advising on ways to perform job specific tasks as well as on navigating the corporate environment.

While these two forms of corporate training are foundational and remain essential parts of ongoing training activities, self-development emerges as a close second. With self-development, staff members take control of their own learning, having decision making authority over the pace and topics. This training, although usually external via enrollment in educational programs, webinars or professional conferences, is highly beneficial for the organization as it correlates with high employee engagement and motivation. When combined with computer-aided training, it delivers low cost, widely available developmental opportunities ranging from basic safety, security and privacy to career development such as leadership training or professional communication skills.

Career development revolves around the creation of a company specific career roadmap strategy for promotion readiness and succession planning. The goal is to develop management and communication skills necessary for middle or high level leadership roles. While one aim is to help future leaders understand the company's strategic plans and prepare them to lead organizational change within their teams, the other is to satisfy personal professional goals such as promotion and career advancement (Reference for Business, 2012). Important considerations include the development of interpersonal and teambuilding skills, understanding the management process, and quality improvement. Teams are made up by professionals with specific skills, abilities and personalities. Their ability to work together effectively and produce high quality output requires specific leadership skills so that members better understand each other's roles, strengths and weaknesses and are better prepared to adapt to expectations amidst changing workplace conditions. Remaining motivated and energized can be challenging so teambuilding is usually an integral part of leadership training and development.

As we can see, the evolution of corporate training is very similar to the progress of career and technical education. Vocational education, which at first was concerned almost exclusively with specific job skills known as competencies is closely related to on-the-job training, originally the only means of developing new workers. The demands of the economy and the society pressured both into drastic changes to now include, in addition to job skills and competencies, core abilities as well. Today's career and technical education as well corporate training and development teaches participants to communicate, think, learn and continually adapt to changing conditions rather than simply providing means to perform certain clearly defined jobs. Training participants have to leave programs as fully engaged, willing participants in organizational transitions and as life-long learners who embrace and adapt quickly to change. In consequence, the term of talent development is increasingly popular due to its more comprehensive approach including traditional training as well as career pathing. According to Wikipedia.org "talent development refers to an organization's ability to align strategic training and career opportunities for employees" (Wikipedia, 2012). This approach fits modern organizational structures better as it allows flexible movement of staff between corporate roles and projects. Flattening organizational charts require broad employee knowledge and skills, allowing the company to easily adapt to changing demands. Changes are fast and ongoing, bringing along perpetual job and career moves that need preparation and adaptation. Global competition requires that staff and company alike embrace lifelong learning and take responsibility for having the skills and knowledge necessary for long-term success. Careers become, in essence, a sequence of learning and relearning skills and transferring knowledge from the classroom to the workplace becomes integral part of strategic leadership.

Thus, talent development evolves into organizational development which concerns strategy development, problem solving, mentoring and career counseling in addition to traditional facilitation. Trainers work closely with business management to accomplish the common purpose of implementing organizational change. They play an important role in responding to and anticipating organizational demands, in identifying the skills an organization and its workers need and helping learners choose between a wide range of learning options. As O'Connor et al. (2002) state, "the decisions regarding workforce training affect the entire organization and the training professional plays a crucial role ensuring that the organization has the workforce it needs".

Organizational development, the "process by which behavioral science knowledge and practices are used to help organizations achieve greater effectiveness, including improved quality of work life and increased productivity" (Clark, 2012) has come to a new stage. Societal changes lead to training targeting priority areas like sales and service, communications, teamwork and leadership development amidst the existing realities of social media, cloud computing and readily accessible online information. As Goodman (2011) states, "at this critical point training and development departments must be part of the global and strategic mission of the organization, define and develop the next generation of leaders and be better than Google". To achieve this goal, corporate training must be solidly grounded in adult learning theory. Workplace learning cannot be effective without such approach because adults don't learn or exist in a vacuum. Their specific situation, existing experiences and backgrounds are virally interconnected with the training outcomes and overall effectiveness of instruction. "If we attempt to teach an adult in the workplace, then such learning should be reinforced with new learning. Additionally, a workplace learning situation must be followed up with a recognition of prior experiences that relates to what the learner may already know" (Boone, 1985).

Organizational development is proactive and overarching, aiming to change future company structures and attitudes in their entirety. It is "the framework for a change process designed to lead to desirable positive impact to all stakeholders and the environment" (Wikipedia, 2012). Organizational development, when functioning at its best, diagnoses and solves for new demands. Much like lifelong learning in career and technical education producing adaptable and flexible employees, organizational development "leaves the client organization with a set of tools, behaviors, attitudes, and an action plan with which to monitor its own state of health and to take corrective steps toward its own renewal and development" (Wikipedia, 2012). Similar to career and technical education where educational goals are achieved via peer collaboration and knowledge sharing, the goal of organizational development is accomplished via interventions that allow staff to complete organizational tasks and work more collaboratively with each other. Team functioning and leadership are essential in this setting, leading to desired organizational outcomes (Clark, 2012).

In conclusion, similar to career and technical education, the mission of training and development is to develop the company's human resources within the context of changing demographics and flexible market demands. Training and development's role coincides with career and technical education's goal of graduating workers ready to contribute to society with their expert level, transferable skills and knowledge. Training provides skills and competencies for current or future positions and helps employees realize their career and life goals (O'Connor, 2002). Student engagement and motivation are key to corporate training as they relate to the individual's success in achieving their maximum potential and to worker satisfaction and retention directly affecting the bottom line.

1.3. Personality Traits that Influence Educator and Trainer Instructional Practices

The discussion thus far on career and technical education and corporate training drew a parallel of the two types of instruction showing that, although different in their setting and specifics, they aim the same ultimate goal of leading people to successful, long-term careers. It looked at the fact that learners and employees coming to career and technical education and training sessions have diverse needs, goals, experiences and entry skill sets. Naturally, teaching such a diverse audience requires a mix of effective strategies and diverse content. Career and technical participants and facilitators alike have dominant personality styles that affect the way they act and react to situations and surroundings. As group leaders, career and technical educators and trainers impact the learning outcomes and the success of their followers by the way they instruct and manage the classroom. Thus understanding the manager versus the leadership roles as well as the facilitator's preference for one versus another is essential for accomplishing the goals of career and technical education and training.

Management is defined as "the organization and coordination of activities in order to achieve a defined goal" (Oxford, 1998). It entails the practice of managing, supervising and controlling processes or a team's actions to accomplish organizational or educational objectives. For management to be effective, a system or plan is needed to help guide those following; managers being reliant on policies to control the group's outcomes. In business, these rules and standards materialize in operating policies and procedures and employee manuals while career and technical educators set the terms and conditions of the educational contract via the policies and procedures section of the course syllabi.

Leadership, on the other hand, "is a process whereby an individual influences a group of individuals to achieve a common goal." According to Northouse (2007), leadership is "a process by which a person influences others to accomplish an objective and directs the organization in a way that makes it more cohesive and coherent". Leadership is based on effective communication, knowing the group members and setting a good example while management implies having the authority to set rules and make decisions in a supervisory position. Rather, leaders "make the followers want to achieve high goals rather than simply bossing people around (Rowe, 2007). Effective leaders are respected due to their hard work, expertise, and ethical behavior and they convey a sense of vision and common goal. As stated by D.R. Clark (2012) "in a nutshell, you must be trustworthy and you have to be able to communicate a vision of where the organization needs to go".

Over the years, research analyzed the characteristics and meaning of the different personalities looking for patterns in the way people act when accomplishing tasks or making decisions or working in a group setting (Ward, 2012). One widely utilized personality styles model is developed in collaboration with expert Kate Ward and published in the book entitled Personality Styles at Work, was administered to more than a million individuals with remarkable results. The model works with the direct, the spirited, the considerate and the systematic personality styles. The direct style is described as highly assertive and low in expressiveness; the spirited style is represented by high assertiveness and high expressiveness; the considerate style is characterized by low assertiveness and high expressiveness while the systematic style is low in assertiveness and high in expressiveness. These styles, combinations of dominant traits, are useful in explaining and predicting behaviors, each being a composite of strengths and weaknesses in the way people approach tasks and relate to those around them. They investigate one's preference for introversion versus extroversion, thinking versus feeling and perceiving versus judging. The strength of one's preference for one category versus the other determines the way they are perceived and consequently accepted or not by other people. For example, while most people appreciate a moderate extrovert in social situations to break an awkward silence, many will also resent someone constantly interrupting others. Also, while most of us are naturally drawn to considerate individuals with an ear always open for others' problems, many get somewhat uncomfortable and perhaps frustrated when unable to connect to an extreme introvert who never initiates interactions (Pearlman, 2011). This means that individuals have preferred ways to act in different situations and in their relationships with other people. For example, someone with a direct style is likely to use short, straightforward phrases and may interrupt others in a conversation while a person with considerate style will wait and engage in active listening. Someone with a spirited style may get excited about events and new projects and talk with high emotional content when faced with problems. Systematic individuals may be seen as even keyed and reliable, making careful decisions based on systematically collected, quality data. Some people have a preference for accomplished projects such as completed degrees and appreciate a clear roadmap of action items to get done while others see no value in completing a degree for the sake of achieving a goal. Rather, they learn for the sake of learning or do for the pleasure of experiencing, simply enjoying the process. Individuals with the considerate preference make sure that everyone is on board with whatever the learner group or team is doing. Similarly, someone systematic easily fits the pieces of the puzzle together into a detailed strategic plan.

The extent in which someone's style is dominant in their actions and reactions determines how they are perceived by others. When a style or strength is overused, it potentially turns into a weakness because they prevent persons from understanding, relating to and accepting the actions or reactions of those around them (Kaplan, 2012). For example, while a direct person's natural, moderate assertiveness is a plus in decision making roles, when extreme due to stress it is likely to be perceived as dictatorial or judgmental and, being seemingly superficial, it will be seen as an area of improvement by others. Similarly, a systematic person, whose natural preference involves well founded research based decisions, under stress or pressure is likely to react by avoiding decision making altogether, the reaction being potentially perceived as a weakness. Yet another example is a considerate who, usually well liked and accepted within the team for building relationships, is perceived as a pushover when avoiding to express opinions or personal preferences in decision making situations.

The context of personality style preference is not the ultimate goal of this text nor does it intend to sound comprehensive in nature. Nonetheless, understanding why a person behaves a certain way helps career and technical instructors and corporate trainers predict their own and others' reactions to classroom and team situations. People act and react according to their behavioral styles and naturally dominant personality traits. Instructors teach, organize the classroom, communicate and give feedback in a way that feels natural to them. For example, systematic teachers and trainers have clearly set classroom rules and well structured course content. Alternatively, spirited instructors may not have the details of an assignment defined and communicated. In general, proactive class communication leads to positive outcomes. However, depending on the teacher's style, these announcements may be more or less detailed and more or less regular. A less organized facilitator may reach out to learners in an unplanned, unpredictable manner without realizing the confusion this can cause. An instructor with a direct or systematic style will be more likely to use black and white verbiage clearly stating the expectations and deadlines, while a perceiver may be more open to adjustments and flexibility. Also, leadership and strategy building come more naturally to spirited and direct individuals than to a systematic individual focused on the abstract details of an existing plan. Thus, it is important that career and technical educators and corporate trainers are aware of their strengths and weaknesses, maximizing the natural traits and abilities and reducing weaknesses with corrective actions. The awareness of how the various styles are perceived by others and how they are represented under stress helps instructors communicate more effectively with class participants and react more efficiently to classroom situations.

Individuals are social beings with a need for relationship building and belonging. They satisfy these needs very differently based on who they are. A direct, naturally impatient individual may prefer short, pointed conversations with little room for irrelevant details while perceivers or those with spirited traits may consider small talk an essential part of every encounter with a well-defined role in building mutually beneficial, collaborative relationships. Classroom examples worth mentioning are instructors providing targeted feedback on student assignments at risk of being perceived harsh or swift, compared to educators composing lengthy explanatory paragraphs on what is well done and what needs improvement. Another typical situation relates to the instructor's patience level with classroom interruptions. Here too, a direct style is beneficial to the extent that situations are addressed and solved, yet overuse of style can easily lead to conflict escalation rather than resolution. Instead a perceiver's preference for avoiding direct conflict in classroom management is easily confused with preferential treatment.

Another aspect of personality traits and perceptions relates to facilitator learning styles. Learning styles are preferred ways to gaining new knowledge and a natural inclination to choose a certain course material types. Usually they are referenced as visual, auditory and kinesthetic styles, delineating a preference for images, diagrams and concept maps, lecture and audio recordings, or hands-on activities. Some authors add a minds-on style to individuals who like to analyze and critically evaluate, while others talk about multiple intelligences, taking learning style preferences to a yet more complex level.

What this means to career and technical educators and corporate trainers is that they have a preference for a certain type of course content and delivery mode. If visual learners themselves will be more likely to choose pictures, create graphs, and highlight key terms versus lecturing the majority of the course materials. Creating extensive hands-on activities will, on the other hand, come naturally to a kinesthetic facilitator. Each style will influence the actual materials delivered to students with the actual percentage rate being different based on the educator's own preference. However, instructors and trainers need to teach all of the students. The teacher's responsibility is to build courses with diverse learner needs in mind, making them as universally usable and accessible as possible. So while they may have their own preference, as educators serving every individual participant in the classroom, they must strive to create an even distribution of course materials types, putting in place an objective value system that strives to even the playing field for all students. Even though as individuals they may opt to represent their thoughts and content with images showing essential correlations with visuals, they need to also provide auditory and text-based explanations to cater to auditory and thinker students in the group. Also, rather than talking in abstracts and expecting learners to seamlessly gap the theory-practice divide, they should build in hands-on practical activities for kinesthetic learners who prefer to do things with their hands. In essence, each content delivery type has its own specific benefits and should be built into career and technical education and training instructional design to create a comprehensive course content structure that satisfies diverse learner needs and preferences.

Personality and learning style type inventories are readily available online and are recommended for review. However, rather than stopping at this level, it is a best practice that instructors and trainers complete career tests and reference printed career guides to form a clear picture of personal traits and strengths in the context of instructional goals and career pathing. For example systematic, yet social individuals will thrive in a teacher role and so will a spirited who learns to set boundaries for good classroom management. Also, while creativity and the ability to make decisions are essential for effective delivery and content development, the right combination works better than an overly creative, yet not organized case or extreme decisiveness leading to premature judgment. The results will provide meaningful insight into natural strengths and weaknesses and one's preference for managing the classroom business or to provide leadership via career and technical education best practices or strategies.

When we take into account the fact that in a classroom community or work group all members have individual personality and learning styles, face diverse work and life realities and react differently to stress, the leader's self awareness of his or her own communications and reactions cannot be over emphasized. All things people do are perceived and evaluated in a slightly different way by each student depending on their own learning and personality traits and background experiences. At the basic level, when instructing with hands-on activities and illustrations, facilitators help kinesthetic and visual learners and when lecturing, they cater to auditory participants. Beyond this, however, there is a whole other dimension of perceptions and interpersonal communications. Communication is a two way process in which the initiator has a specific intention and relies on his or her own existing knowledge and background experience. The receiver of the message is the target audience, each individual filtering through the meaning of the words received based on their own knowledge system, values and purposes. Even without considering the distortions due to noises on the communication pathway the words will be understood differently by each member of the target audience. With this, come numerous possibilities for misunderstanding and conflict, making awareness of perceptions and details an essential concept for career and technical educators and corporate trainers.

Personality styles and context act as filters, modifying the intended meaning of the messages sent into something unintended and often unexpected. As educators work with large participant groups they can reduce the possibility of being misunderstood by proactively planning on the fact of learner diversity. Crafting course content and announcements according to this guideline will increase the effectiveness of their communications. Similar to utilizing mixed course materials incorporating visuals, audio and hands-on elements even if their own individual preference leans toward just one of them, they also need to modify their communications from what first instincts say to something that is likely to achieve the desired outcome. Thus, even when their natural tendency may dictate a clear cut, short message containing facts only, they should strive to send announcements that also include soft talk and encouragement that students perceive as friendly, collaborative and having positive intent. They may also need to modify some of their automatic traits and tendencies to better fit the audience. For spirited individuals, their natural tendency may be to cheerlead and motivate the participants along their educational journey, but they also need to assume direct or systematic roles to create an environment that incents success via its clear structure, relevant content and logical flow. Similarly, even if one's tendency is to emphasize systemic setup and strictly informational communications, they need to also consider the leadership side of teaching and build in regular affective engagement elements that encourage student effort.

This book aims to show that career and technical instruction and corporate training are very similar even though they operate in two different contexts. They each have two sides, business management and leadership, which, when masterfully incorporated into the final instructional design, lead to effective classrooms and training sessions. The operational best practices resemble managerial duties while the strategies relate to the leadership side of instruction. Their combination, placed into a comprehensive course structure, seamlessly guides participants toward long term career success. Thus, the goal of this book is to explain that indifferently of one's natural tendency to focus on management practices or on leadership strategies, the career and technical teachers and corporate trainers have the professional responsibility of using both approaches. This will assure that they create effective classrooms and provide quality education that satisfies the needs and expectations of the participants as well as the economy. The following chapters will look at specific practices and strategies that can be used to realize the goal of effective instruction and participant success by turning strict management into skillful leadership motivating high achievement and academic accomplishment.

Chapter 1 Summary Overview

This chapter looks at the mission and goals of career and technical education and the evolution and role of corporate training. It discusses key instructional approaches including competencies and core abilities, real world applicability, industry standards, academic rigor and critical thinking and lifelong learning. It also provides an overview of the transformation of the corporate training and development function and the role of new hire and on-the-job training. The chapter concludes with a review of basic personality and learning style preferences and the relationship between traits and career and technical instructional outcomes.

Chapter 1 Key Terms

Academic Rigor
Career Readiness
Communication Styles
Corporate Training
Employee Acclimation
On-the-Job Training
Organizational Development
Personality Traits
Practical Applications
Student Success

CHAPTER 2

Practices for Managing Career and Technical Education and Training Instruction

The objectives of this chapter are to lend instructors a basic knowledge of:

- the role of classroom management in achieving instructional outcomes
- effective course syllabi and calendar instructional practices

2.1. The Role of Classroom Management in Achieving Instructional Outcomes

Successful businesses and effective classrooms start out with efficient management. The business of career and technical education and training involves having clearly defined and strategically enforced standard operating practices. In corporate training, new associates need to be acclimated to the organizational vision to safety and privacy rules and customer service standards to establish their long term profitability. In career and technical education, the instructor also needs to set policies and rules because students, like new employees, cannot be expected to understand the expectations without proper training.

In business, the first day on a new job is essential for retention and for maximizing the new employee's satisfaction. Similarly, the first class meeting has foundational role in effective classroom management. Most new hires strive to make a good impression on their first day by arriving on time and wearing professional attire while adult learners come prepared and focused on understanding the expectations of the new environment. They are eager and are fully engaged in getting details such as course goals, topics, and assignments. At the same time, they also are weighing the instructor's authority, knowledge, and professionalism. As adults who build on their existing work experiences and connect classroom behaviors to starting a new job, this first day can be called the window of opportunity for setting ground rules, establishing authority and emotionally engaging the new individual. It is a short interval in which there is a high probability that the expectation will be heard and accepted. In career and technical education, explaining the rules and deadlines while also showing learners the connection between theory and applicability gets good return on the instructor's investment. Similarly, when new staff arrives to the job the first time, trainers or managers have a chance to set limits and expectations as well as to connect with the new employee. In both settings, these initial interactions are natural and expected, reaching the receiver's attention to create meaning and should be carefully designed to achieve a specific goal. It is a best practice to start out strong and set firm rules during the window of opportunity and relax them, if desired, later on, because, as a general rule, it is easier to get easier than the other way around. Classroom management, defined as "the process of ensuring that classrooms run smoothly" (Wolfgang & Glickman, 1986) is, like business operations, three quarters planning and one quarter intervention. As the practice of strong foundations pay off in construction, in business strategy development or personal financial planning it is just as effective in classroom management. A business strategy does not produce desired results if the starting assumptions are erroneous. A classroom which does not have rules will also deviate from accomplishing its mission. In business, the employee's first day on the job is important for long term productivity and work satisfaction. If, upon arrival to the company at the time requested, the new staff needs to wait in vain for awhile prior to anyone welcoming them, the new employee may erroneously assume that the company does not value punctuality. In result, the new associate may be more likely to exhibit tardy behavior. Also, if the on-the-job mentor appointed to the new hire shares negative comments relating to leadership decisions, the new hire is more likely to convey similar information to others, considering it acceptable behavior. First impressions count in both the classroom and the corporate training environment. Once the class moves on, the ability to set ground rules and authority is rapidly reduced and if too much time has passed, the chance of creating an environment of mutual respect and collaboration becomes very slim. According to Moskowitz and Hayman (1976) "once the teacher loses control of their classroom, it becomes increasingly more difficult for them to regain that control". Setting a positive, professional tone at the beginning pays off in positive goodwill in both settings, while lack of professionalism negatively impacts general business operations and instructor reputation alike. Closely related to the issues of motivation, discipline and respect, "rules give concrete direction to ensure that expectations become reality" (Gootman, 2008).

For students, the content learned during a class is a product they purchased and intend to use long-term. Naturally, students don't want to waste valuable financial and time resources on classes that do not fit their needs, wants and expectations. Thinking through students wants, needs and expectations one will soon realize that they are very diverse and it is unrealistic to expect their equal satisfaction. However, one thing that most students have in common is a desire to achieve a credential such as a completed course, a certification or a degree. Also, what students need is lucrative knowledge and skill applicable to the career of their choice. At every educational level, from K-12 to graduate studies, educators must assure that learners leave the classroom with the competency set they aimed to attain. This is what students expect in context of achieving their credential. These requirements can be satisfied only through rigorous academic requirements enforced in a learning environment focused on helping each student strive. Reducing the rigor means setting them up for career failure as they leave lacking the skills and knowledge they need to be successful in the industry setting. Failing them isn't the answer because it does not allow achievement of the credential they need for long term careers. Thus one can see that the ultimate goal and only correct approach is to create effective classrooms that deliver on students' expectations. This can be achieved via successful instructional design, rigorous expectations, communication, and learner engagement strategies. When these elements are present, students are able to fully engage in learning from day one, moving, via structured content and specific formative feedback, toward the goal of course completion. This approach assures that knowledge and skill attainment is regularly evaluated and improved and students achieve their goal of gaining a credential. Only with implementing effective classroom management practices can educators and trainers satisfy student expectations. In fact, without effective classroom management, instructors set students up for failure by creating a confusing environment in which they do not know what is expected of them, why and when. They do not see the reason or outcome of their learning or don't understand how to succeed, de-motivated by trying to follow a cumbersome path. Feelings such as loneliness, anxiety and lack of control emerge and the path of abandoning the educational roadmap becomes tempting. Returning to behaviors and actions they are familiar with even though not socially accepted or economically practical, becomes an easy route to pursue. Educators striving to lead students to success must catch learners' attention in the highly effective window of opportunity. Doing so leads to creation of an effective and motivating learning environment with practical outcomes and clear expectations as, in matter of minutes, students can visualize the goals as well as the roadmap to achievement. By showing the image of success and the heights education can lead them to teachers engage learner dreams and passions and build a clear pathway for them to follow.

To determine the specific elements of starting a classroom on strong terms, one can look to a corporate general new hire training session example. New employees, when attending this type of training, reasonably expect to see a company overview including mission, values and organizational structure, a leadership welcome, a map of locations or merchandise lines, a quick reference for computer and security systems and a listing of key contacts for essential questions. The comprehensive new hire training when combined with positive and effective first day impressions leads to a strong corporate start. The career and technical education setting is not different. If we consider the educational institution a business enterprise, we'll notice that the foundational elements of student success are similar to those needed for long-term employee efficiency. Thus students should be provided with a complete welcome package including specific course documentation. The course syllabi component of the welcome package can be considered a representation of company operating policies and procedures. It is recommended that the welcome package contains a course calendar also, with the role of communicating the course structure and lesson plans. These elements create an intimate, yet realistic connection and create instant educator authority.

Like new employee training which goes beyond job satisfaction and performance to having a risk reduction aspect as well, classroom operations work much the same. For example, a quality procedural overview regarding company pricing, confidentiality and security procedures can protect against future losses and law suits. Setting firm ground rules on essential policies such as tardiness, teamwork, plagiarism, and assignment submissions are going to be foundations of a well run classroom. With the pieces put together correctly and maintained appropriately, the classroom will run like a well oiled machine needing little intervention. Extensive detail in the course syllabi can be rather involved and time consuming. However, just as successful businesses have complex functional structures encompassing finance, operations, marketing and leadership, it is natural that one of the foundations of career and technical education and training classroom management is just as comprehensive. Given its direct link to quality, not giving it the attention it requires discounts the relevance of the next generation's education.

Next the analysis moves onto ground rules and instructor authority. Instructor authority comprises outstanding subject matter expertise and exceptional teaching ability. In career and technical education, the term instructor authority symbolizes a combination of course related knowledge, practical experience and students focus. Like the authority of a medical or financial service provider, learners representing the educational industry customers, seek assurance of high quality services. Customers select their provider based on confidence level, being willing to pay for the services of trusted professionals who adhere to high standards of excellence. Truly, the first and most important factor customers look at is professionalism and expertise. Much like with personal health or finances, they keep the quality of their education forefront, wanting nothing but the best for themselves and their children.

Quality, as a "peculiar and essential character or inherent feature with a high degree of excellence" or the process of "checking goods and services at all stages of production and delivery to assure that they are of a good standard" (Oxford, 1998), regards an extended period of time in which the product or service is expected to function according to specifications. The term can then be related to product life cycle and warranty. Educational quality encompasses excellence in content delivery and assessment during the duration of the course and beyond. Products are expected to function well beyond the expiration of their explicit warranty terms. In fact, if the product does not function properly during or after its warranty, customers will question the overall quality and reputation of the company. Similarly, clients choose financial professionals with great care, fully understanding the importance of quality long term financial planning for achieving financial goals and dreams. Trust is a key element in this decision making as they seek assurance that the advisor has the knowledge and ability to perform the tasks necessary for staying on the chosen financial track. Of course, right next to expertise and professionalism customers also look for a caring attitude and a long term reputation of excellence that encompasses relevant education, experience and positive customer evaluations. In context of service reliability, price and variety these aspects are evaluated in conjunction with the perception that the institution or professional is able to deliver on its promises consistently and reliably, on demand. The provider's ability to perform according to expectations depends on the effectiveness of their management and company rules. These ground rules of operation, encompassing a set of established policies and procedures are key to profitability as a business, without effective practices in regards of sales goals, customer privacy, transaction accuracy and service quality will fail to deliver on its promises.

Ground rules are also relevant for classroom management. An instructor who does not have firm ground rules is set up for failure as quality education is only possible when students are aware and adhere to well defined, well grounded expectations. Students, the educational industry's customers have the right to a safe and effective learning environment. In consequence, instructors have the professional responsibility to set ground rules for the benefit of students who paid for quality services. Ground rules are just as necessary for the classroom effectiveness as operational procedures are for running a profitable business. They are foundational elements of managing the career and technical education classroom as they set the boundaries of what happens in the learning environment. They also define what is meant by professional behavior or tardiness, teamwork, etc. expected of learners. The explanation of the ground rules adds a second layer to the class terms set forth by the course syllabi. They facilitate open discussion and explanation of the policies, putting a personal touch on the dry contractual terms. For example, if the course syllabi states that tardiness or interruptions will not be tolerated, through the ground rules discussion the instructor can specify that an occasional two or three minutes late arrival due to a family or work issue is evaluated differently than ongoing, unexplained tardiness. Similarly, this is good time to explain that cell phones should not ring during class and only in an emergency situation should a student leave the class to answer an incoming call.

Setting the stage with solid ground rules has two main roles. First, they open the doors for two-way communication around specific expectations and why certain rules are necessary and beneficial for individuals and the learning community. Second, ground rules are preventative in nature. They help reinforce the basic policies set forth via the course syllabi, allowing instructors to emphasize selected elements considered very important.

A classroom management rule set forth via the course syllabi and reemphasized via the ground rules resembles the foundational level of a layered cake. The second level can be the essential classroom prevention strategy of general, large group reminders at the time of occurrence of a specific incident. For example, if the teacher notices that a student arrives five minutes late twice, the ground rule relating to acceptable and unacceptable tardiness should be revisited in class with the whole group. It is a best practice to follow up the in person group discussion with a written announcement for traceability. No specific identifiers should be mentioned;

the goal here is to express that the incident was acknowledged and needs to be redressed. A similar corporate training case relates to a new employee who arrives late to work twice in the first two week period. It is advisable and considered good business practice that the employee's trainer or mentor first discusses work arrival time expectations in a staff meeting in a generic manner and only if the incident surfaces again proceeds to have a private conversation with the worker. The same preventative procedure, is also applicable to classroom management, as all rules and reminders that happen prior to the private conversation are considered proactive prevention strategies.

Indeed, classroom management is two thirds prevention and one third intervention. In other words, well laid foundations lead to a strong, healthy learning environment that needs few, if any, repairs along the way. After solid basis, maintenance comes next in importance. In the educational setting, this means holding students accountable to the set rules and expectations. If prevention strategies are deemed insufficient a private follow-up discussion needs to be held to bring the individual student's attention to their deviation from expected. It is a best practice to discuss what is acceptable and what is not to reinstate, now firmly and with specific detail, what needs to change. The discussion should be followed by a written correspondence as to provide traceable evidence of the instructor's proactive prevention efforts.

In business, this discussion is usually called coaching. In normal circumstances coaching is to take a sandwich approach, pointing out two or three positives as well as specifics improvement opportunities. In the case of employee performance management strategy, the coaching conversation is be very specific to the incident and the areas of improvement. If the employee redresses behavior according to expectations the performance management stops. If not, appropriate next steps need to be taken as prescribed by the company's performance management policy. Just like in education, the steps taken are to be documented to provide evidence of actions and intentions.

If the private conversation is not successful in improving behaviors, the next step in classroom and performance management is to be taken. In career and technical education this entails submission of a counseling or an advising referral while for business, this means a written performance documentation submitted to human resources. The electronic submission processes in place at most educational institutions do not require the parties to physically sign rather they send an automated email notification to both parties stating that a submission was processed. The outcome is the same in both settings, the individual being held accountable and yet provided with another opportunity to improve or get appropriate help.

An important topic is the severity level of the inappropriate behavior exhibited. For example, being five minutes late once is different than having ongoing side conversations that distract the rest of the group or coming to class under the influence of drugs or alcohol. Naturally, severe incidents such as these necessitate immediate intervention. In business and education alike, repeated, unexplained tardiness or any substance abuse draws immediate dismissal. The step of reminders is skipped, classroom and performance management starting at a heightened level of severity. A written notification is forwarded to the appropriate institutional entity as well as to the individual in question with clear expectations on immediate improvement. It is a best practice to use available security personnel to avoid risk of personal injury and to assure protection of the entire class or work group. Cases of alcohol or drug abuse or aggressive behaviors lead to immediate security referrals to escort the student off the premises and home safely. This procedure, in addition to creating a safe environment for everyone, also assures that the student in question gets the help or treatment needed to get well. From this perspective, the measure is preventative on large scale as when learners see a peer escorted away by security officers or law enforcement will be less likely to overstep their boundaries by displaying inappropriate behavior themselves.

The next step is to delineate classroom interruptions from disruptive behavior. Classroom interruption examples include repeated tardiness, randomly ringing cell phones, occasional side conversations and confrontational behavior and negative comments. Interruptions, although when isolated may not seem too severe, become classroom management issues when repeated or forming a pattern. Handling individual situations while isolated is a best practice to stop the formation of the snowball of complex issues. After setting solid, safe foundations via ground rules, the next step is to intervene effectively when the interruptions appear or at the first signs of pattern formation. It is important to note that patterns can involve repeat offenders or multiple students. Basically, if the snowball starts to roll, it grows quickly by picking up additional snow from all sides. In fact, even in everyday verbiage the snowball effect metaphor depicts exponential growth of something that starts out small and then explodes out of proportion, faster than expected.

For instructors this means that interruptions need to be corrected right away to avoid a large scale classroom management disaster. In the business setting the importance of solid boundaries and holding individuals accountable is represented in performance management and cannot be over emphasized. In companies interrupters are employees who arrive just five minutes late every day or who avoid unpleasant office tasks causing others to have to pick up the slack repeatedly. Another example of workplace interruptions are team members who consistently leave on or return from break five or ten minutes late, causing delays in the appointments or meetings of others. Even though some of these instances may seem petty, they have a significant role in the creation of an unpleasant work environment and feelings of resentment towards management. Once mistrust sets in the group is likely to question every decision leaders make. Another serious interruption type is the student or employee whose negative attitude is vocalized by comments like "this is not going to work" or "this topic is irrelevant, we'll never use this in real-life". These comments will undermine the leader's authority and create a negative, offensive and volatile classroom atmosphere which can easily escalate from the individual to the entire group. Suddenly instructors or managers are faced with a large scale issue affecting the group's performance and collective membership. The snowball effect is, once again, in full action. Stopping it at this point takes harsh actions such as security intervention or performance write-ups which create a general sense of fear instead of buy-in and engagement. Same is true for the classroom; if things get out of control, the instructor gets a reputation of ineffectiveness and even after the situation is redressed, a general negative atmosphere replaces that of a collaborative learning community.

Educational institutions are perhaps even more affected by such incidents than privately owned companies because while employees are bound by strict confidentiality rules, students are prone to openly sharing their experiences with others. Thus the snowball of negative goodwill starts rolling, magnifying the original issue out of proportion and adversely affecting the institutions' reputation. Disruptive classroom behaviors are elevated intensity interruptions or those that form a pattern. When instructors have a good handle on classroom management with firm ground rules and good prevention plans in place, most interruptions never escalate to the disruptive level. Examples of disruptive behaviors are learners walking out of the classroom slamming the door behind or students who arrive or leave mid-class. Such situations warrant immediate intervention, the educator needing to curb such unacceptable actions in their tracks. In fact, a lack of intervention leads to lost authority and respect from the learner group.

Disruptive behaviors such as cursing, sexual or biased verbiage or an intentionally challenging attitude also need immediate attention. Other examples of situations that warrant immediate intervention are cases of family or work distress, fear, loss, anger and anxiety. The timing and types of the most appropriate response depends on the degree of violence displayed or the intensity of risk involved. Teachers have two basic routes to choose from. One involves an immediate security referral to safely remove the student from campus premises. This type of intervention, although harsh, is effective for maintaining a safe learning environment for all students. Calling campus security personnel is recommended because it creates a sense of a effective management and safety while also providing appropriate assistance to the learner in question. The student who comes to class under the influence of alcohol or drugs is not only a danger to the group but also to themselves. It is recommended to let trained security officers handle the potentially explosive situation using specialty emotion reduction techniques and established campus procedures. This approach eliminates the risk of instructors mishandling the situation, involuntarily increasing it in intensity. Colleges have clear procedures for escorting disruptive students safely to their final destination rather than just outside the building, avoiding the possibility of further injury to self or others.

Nonetheless, not all disruptive students fit the definition of violent, loud or aggressive. Some learners are quite the opposite, displaying passive-aggressive behaviors or struggling with depression or anxiety disorders. Such students often are extremely quiet and don't engage in activities or with the group. Some may disclose their fear of failure, anxiety or stress due to a lost job or loved one. They, although need immediate help, generally require no security personnel intervention. More so, they are at risk of being unsuccessful in the classroom and life alike due to their disturbed emotional state or perhaps an undiagnosed disability. In these cases, instructors have the responsibility to extend effective, targeted help as, if they go unnoticed, they can turn into immediate threats to themselves or others. These students need educators perhaps even more readily than the others because openly aggressive or addicted learners first need specialized psychological and medical attention and only afterwards can they receive instructional coaching. On the contrary, sad, anxious, lonesome and depressed learners may actually be much more open to the hope and help teachers can provide via empathetic listening and caring guidance. These students need advice and support. Often all it takes is to simply be there and let them think through their issues and come to their own conclusions. Other times it entails asking additional probing questions to determine what kind of support would be best suited for them. Very often, instructors are able to make suggestions just based on their own personal life experience or specialized expertise. Other times it may be advisable to forward the case to a specialized resource center or professional service. For example, if a student is struggling with work related anxiety or job loss, educators may be in position to help with job search or coping strategies. Yet in other cases, their best intervention may be to simply refer the learner to a community resource or student support center. It is important to actually take the active step of submitting the referral to counseling service to make the essential first connection between the distressed learner and the trained professional. Proactivity is key here too, providing the individual with the help they need and effectively reducing the risk of escalation from a classroom management standpoint.

In order for instructors to be effective as guides and mentors, they need to fully understand the distinction between the counseling and advising as student support services. Counseling and advising, although often housed in the same department, work with different aspects of student success and are two distinct levels of assistance available to post-secondary learners. Advising usually works with academics related issues such as course sequencing, course load, financial aid and degree attainment, attendance, first year orientations, etc. Examples of classroom issues warranting an advising referral are cases of repeated tardiness or late homework. Students with special needs due to diverse backgrounds or medically documented disabilities would also utilize advising support services. Counselors, on the other hand, work with behavioral and mental health conditions such as anxiety, depression, alcohol and substance abuse, etc. When noticing the signs of these conditions, teachers need to intervene by referring the learner to college or community service providers to receive specialized help at an affordable cost.

The confusion comes in when instructors do not know exactly what causes the student's distress or are not in position to correctly determine the most appropriate type of assistance necessary. In response many institutions implemented a single point of contact automated referral system for instructors to submit their observations, which then forwards the instances received to the most appropriate specialized personnel. This approach is highly recommended as repeat names or continuing cases can be routed to the same counselor for better targeted and more personalized assistance.

After the discussion on career and technical education interruptions and disruptive behaviors, it is time to analyze the industry training aspects as well. In the business setting, isolated cases of tardiness or an occasional negative comment seems like something that hardly qualifies to be a disruption. However, the perspective of managers not holding their team accountable to expectations and company policies leads to their categorization as disruptive business behavior. To understand it better, it is helpful to use a financial service example. An employee being late once due to a sick child and the second time due to car trouble, although seemingly having plausible explanations, is in actuality an interrupter of safe and effective operations. In the context of the financial center management where two employees need to be present for opening the tardiness affects the institution's overall safety and possibly its reputation. The safety aspect elevates the employee's repeated behavior from an interruption to disruptive behavior as, due to the circumstances impossible to predict, the situation could easily get out of control or result in robbery and risk of harm.

Another workplace disruption is using loud or foul language with peers, supervisors or customers, or simply avoiding following company rules. Depending on their severity, these cases may warrant immediate dismissal from employment while may lead to less visible but still assertive measures such as being written up for inappropriate conduct. Similar to the measures used in the post-secondary education, trainers and mentors may need to involve law enforcement for effective and safe management of explosive workplace situations. For example, if an employee is under the influence of alcohol or illegal substances, safe removal from the company's premises as well assuring their safe arrival to their home destination are essential and are best accomplished with the help of local law enforcement. Of course, should the company have its own security personnel on staff at the site in question using them will take priority to reduce cost and reputational damage. The most important thing is to protect staff members and customers from physical harm and to prevent negative goodwill formation, both routes being much more advisable than the non-intervention alternative.

Another damaging workplace situation is use of biased or discriminatory language, sexual harassment or defamation. These qualify for the disruptive behavior definition because they prevent safe and effective long-term company operations. Rarely open and visible, this type of conduct manifests in form of insubordination and harmful gossip targeting peers or management, creating an ineffective and unpleasant environment. Preventive workplace management, including sound ground rules and reinforced expectations, effectively prevents most issues. Open communication and regular, ongoing relationship building with staff members also opens the door for fully understanding the cause and source of such problems. But, if disruptions do arise, effective and immediate interventions are necessary.

Yet another aspect relates to emotional issues employees may be struggling with such as depression, anxiety, divorce or grieving or concerns relating to sick children, elderly parent care or single parenting. These situations are usually part of ongoing management but they can easily get out of control and become disruptive. Examples are multiple team members being out on long or short term medical leave or requesting flexibility due to school age children or aging parents. These cases utilize the manager's guidance and mentoring skills, helping staff cope by contacting workplace or community services or simply providing emotional support via empathetic listening. Although in the case of business operations, referrals are not allowed on behalf of employees, continuous monitoring of progress can assure that the employees get the help needed and that the situation does not escalate from a workplace management standpoint.

Our discussion of the career and technical education classroom and training would not be complete without taking a close look at special needs accommodations. Legislation defines special needs students as "those with special educational needs relating to additional resources or services to support their education" (OECD, 2012) and requires educational institutions as well as businesses to provide certain reasonable accommodations for documented disabilities. This is a broad definition that looks at disabilities very differently depending on the context. While in K-12 education there is a heavy emphasis on mandated accommodations, in the post-secondary setting the requirements are less stringent or inclusive. However, the meaning of the term is similar for post-secondary career and technical education and corporate training. In these settings, while it is considered reasonable accommodation to provide means for the disabled to access their workplace and classroom safely and to create equal access to academic achievement or job performance, the legislator does not ask that instructors or companies reduce or alter the level of rigor or quality expected from these individuals. Also, while adjusting surroundings to some extent to allow disabled individuals to study or work is considered good business practice, the law does not mandate or enforce specific details. For example, if a production facility uses business machines with green and red decision making indicators, the laws does not make the company or the post-secondary career and technical educational institution change or alter its shop equipment or lab setup. Also, although usually provided by schools, making all textbooks and equipment manuals available in audio and Braille format is not mandated because these accommodations may not be available in the regular industry setting where the individual is to function and succeed. The legislator's goal is to help participants with disabilities succeed in the real world industry environment, similar to allowing clear posting of job task related physical, visual or auditory ability needs.

Beyond physical disabilities, educators and managers alike are sometimes faced with invisible disabilities that can be extremely hard to identify. Both workplace leaders and teachers need to investigate in good faith the reasons for the unusual or unexplained behaviors and provide reasonable accommodations in context of regular classroom or business operations. It is important to note that post-secondary career and technical schools and companies need not to be concerned with accommodating alleged, but not documented disabilities. The individuals in question are accountable to disclose and provide medical proof of their condition as well as the specific accommodations they need. Educational institutions and businesses evaluate the possibility of creating cost effective and reasonable conditions for assisting the individual and, if possible, take the steps to create the accommodations within the shortest timeframe possible. The legislature provides a high degree of freedom for institutions in determining what is reasonable and cost effective in their specific situations, mandating the overall goal rather than the specific details. This is explained by the mission and goals of career and technical education and corporate training relating to preparation for the real world of work. Even though the ultimate purpose of any educational institution is always the success of every learner, career and technical education is set in context of practical industry connections. An example is a truck driver education program or training series in which the institution is not required to enroll color blind or deaf persons because career success of these individual is not possible in the real world trucking industry. If the appropriate special accommodations are provided in the chosen career field, such as trucks with special controls, then the educational institutions needs to make them available. If the accommodations are disproportionately high cost or not reasonable by industry standards, the legislation does not force the school or the company to accept the individual's request for enrollment or extend a hiring offer.

This point is most appropriate for discussing Universal Design for Learning which is "the design of instructional materials and activities to make the curriculum equally accessible and appropriately challenging for individuals of differing ability, background, and learning styles" (UDLC, 2012). The term was originally developed for architectural design to create structures that allow diverse users to utilize them to a maximum extent possible. Key elements relate to cost effectiveness and proactive planning for diverse user needs in all aspects of work and life. A straightforward example is implementation of alternative access ways to buildings and development of safe toys for young children. In education, universal design is incorporated into the institution's real estate such as wheel chair accessible classrooms, facilities and student services, web based and large print publications, etc. Universal design should also be part of course design meaning that instructors are to plan content and delivery methods for a wide variety of learner needs. A simple example is using mixed media materials such as text, video, audio and visuals or avoiding the use of red and green text in assessments. Other examples relate to consistent use of font style and size, video captioning and alternative text for audio files incorporated in instruction.

Companies also need to be mindful of universal design principles. Although many details relating to the accessibility of corporate services is in form of legislative recommendations rather than mandated specifics, businesses benefit from implementing easy to use, cost effective features that serve their entire employee and customer base. For example it is a best practice to caption the company president's video welcome message to assist those with hearing deficiencies and those whose first language is not English. This approach creates a positive and inclusive work atmosphere due to being culturally responsive as well as welcoming. The corporate training environment, much like the career and technical education classroom, needs to go beyond the surface of disability accommodations, creating an inclusive learning environment that is effective regardless of one's age, abilities or background. The goal of corporate training is high performance, making trainers just as accountable for the effectiveness of their professional development offerings as educators are in the classrooms. Naturally so, as career and technical education is, in reality, an extension of workplace development, serving a highly diverse audience. Proper preplanning leads to effective instruction in both contexts.

Another aspect of the business of career and technical education and training is sharing experiences, knowledge and information. Most adults are perpetually connected via electronic communication tools and cloud-based applications. A shift has happened, changing the social mindset from a private and copyrighted to a more socially open, value-added collaboration in education and industry alike. The norm changed to ongoing display of benefits brought to customers, community and society in general. In consequence, our students and employees are encouraged to share and stay in touch with clients and peers inside and outside the walls of the organization. Educators encourage the collaborative brain approach in classroom activities and group work, yet they also need to make certain that learners leave with appropriate skills and knowledge outcomes. To accomplish this, they build in verification strategies that assure that students have completed their homework and assignments themselves and did not, in fact, collaborate with or borrow someone else's work. While instructors want them to research relevant resources, they also forbid them to plagiarize by directly quoting existing sources. In this controversy, it is natural that students get confused and may feel unsure about when to share and when not to share and when is collaboration appropriate. In addition, there is a serious generational gap in the approach to individuality versus teamwork, personal ownership or shared resources. Same is true for the corporate training environment. Employees are encouraged to stay in touch with customers and develop long term, trust based relationships. Companies provide tools for researching the background profile of targeted clients and train their staff on using electronic and social communication tools effectively. And yet, employers also expect employees to adhere to best business practices related to customer privacy and trade secrets, business strategies, safety and confidentiality. Naturally, staff can get confused as to what can or cannot be done. Employees struggle with what communication is considered beneficial relationship-building and what is harmful gossip with repercussions.

The topic presents itself in a very similar way in both contexts. Luckily, the applicable best practice solution is also similar in its logic, coming down to setting clear expectations from start. In education this entails development of a comprehensive course syllabi, establishment of firm ground rules and creation of effective intervention procedures. In corporate training, it relates to establishing operational procedures and policies that guide everyday employee activities and decisions. In both contexts the rules or procedures serve their purpose only if they are followed consistently. It may be true that in post secondary education the risk of sharing and plagiarism cannot be completely eliminated, it can be reduced by good planning and institutional design. In corporate training and development, as well, the risk of employees misusing freedoms or tools cannot be avoided entirely. It can, however, be managed by effective operational and performance management practices such as those discussed earlier.

It is important to look at inappropriate sharing of information in the classroom and the workplace settings from another perspective as well. The discussion showed that learners and employees can get easily confused and rightfully so. Sharing is, on the other hand, a highly beneficial teaching strategy and business practice that is to be incorporated in every day operations to maximize its benefits. Keeping this in mind, instructional practices need to incorporate the possibility of remediation for honest mistakes. In the educational context instructors need to be proactive in assuring that learners submit their own homework, but they should also consider rework and resubmission for first-time offenders. For repeat offenses, a layered approach incorporating established institutional procedures should be used. In companies, as well, even if inappropriate information sharing caries reputational risk, employees are to be given an opportunity to redress their behavior prior to being put through intensive performance management.

The final topic of discussion regarding best operational practices in career and technical education and corporate training is assessment management. Assessments are, in a nutshell, different modalities of demonstrating knowledge or skill relating to a certain topic. "Classroom assessment is both a teaching approach and a set of techniques" (NTLF, 2012). According to Angelo and Cross (1993), "classroom assessment is a simple method faculty can use to collect feedback, early and often, on how well their students are learning what they are being taught" different assessment types existing within the broad definition.

Career and technical education's approach to assessing knowledge is as practical as all the other aspects of instruction. It includes formative and summative evaluation techniques, each having a distinct role in assuring that learners gain the skill or knowledge intended. Summative assessments, defined as "a means to gauge, at a particular point in time, student learning relative to content standards" (AMLE, 2012) are evaluation techniques that allow instructors to gauge the breadth and depth of learner knowledge, including content volume and key correlations. The goal of summative evaluation is to provide feedback on the overall quality of the learning outcome. This assessment type is generally used for mid-term or final evaluation and can take the form of tests or research projects or portfolios. While specific project aspects and details are evaluated just the same as with formative assessments, the main difference relates to the fact that the strategy aims to evaluate comprehensive end knowledge. Once the submission is complete, the grade awarded is final and non-alterable. As a result, it is best to utilize summative assessments as high value grading items at mid-term or at course end, providing an incentive for learners to buy into their preparation.

In contrast, formative evaluation "provides the information needed to adjust the teaching and learning while they are happening, informing both the teacher and the students about student understanding at a point when timely adjustments can be made" (AMLE, 2012), serving the purpose of continuous improvement of learner skills and knowledge during the course. Formative assessment tools range from first drafts on essays and self-test online quizzes to ongoing feedback on projects. Formative evaluation can be used, as instructor feedback tools as well as opportunities for peer or external critiques. In fact, peer and role model feedback is, in many cases, more effective than instructor evaluation because it increases student buy-in, students being more likely to accept and incorporate the suggestions into their final projects. Instructor generated formative feedback benefits both parties by allowing ongoing insight into the development of students' skills and knowledge as well as into the instructor's expectations. By giving and receiving specific, targeted feedback, both sides profit from clarification or additional attention. In education and business alike, it is good coaching practice to pinpoint specific areas that need improvement while also acknowledging areas that are done well already, to enhance motivation to achieve high level goals and increase final outcome quality.

In corporate training assessments are considered just as much a key practice as in education, although they are done using different tools. In the company setting, employees are evaluated in conformity with their level of responsibility using preset performance management procedures. These processes are usually structured into layers of evaluations with varied frequency and importance. A typical check-in is done weekly and coaching sessions, representing more in depths discussions, are generally scheduled monthly. Finally, the capstone of corporate performance management, the yearly performance review incorporates a list of accomplishments and action items for improvement. In this setting, formative feedback is represented by ongoing coaching conversations while summative evaluations are done by means of the yearly performance review document.

Corporate performance assessments mimic the procedure and value of their career and technical education counterparts. Instructional formative evaluations are scattered across the continuum of the course being graded with relatively low point values, while summative assessments are placed at mid-term or course end, being awarded significant value. Similarly, in corporate employee development, coaching discussions are regular and low key, while the major performance evaluation is scheduled once a year, having serious significance. The document incorporates aspects discussed during coaching sessions and is considered key for internal promotions and pay raises. Naturally, this leads to high employee engagement and motivation, the performance review being one of the most important and respected tools in a leader's toolbox. Similarly, the final examination or capstone project is the highest stake item in the career and technical education instructor's control.

This concludes our discussion on the best business practices used in career and technical education and training, allowing us to take a step forward and examine strategies that allow facilitators to implement operations and management processes with masterful skill, leading their teams to success.

2.2. Effective Course Syllabi and Calendar Instructional Practices

The keys to career and technical education instructional effectiveness are the course syllabus and the course calendar. The practices relating to these two foundational elements are building blocks of basic classroom management and are essential for creating an environment conducive to learning. It is good practice to start with an overview of the policies and expectations combined with time for questions and mutual commitments. A short introduction to the instructor's professional expertise and educational background can be done at this time. The level of general preparedness including course details, classroom technologies, appearance, and approachability are all best practices for the first class meeting.

Most educational institutions have a designated course syllabi template instructors can use. It is recommended that new instructors use the generic template and ask for existing samples from peers because starting from scratch is counter-indicated due to inherent complexity and potential risks. It is a best practice that novice educators start by adapting the general model to their specific needs, keeping intact institutional practices and policies while adding particulars to the flexible areas such as the course calendar or grading points. Once in possession of the institution's general course syllabi template, educators can than add and remove elements or adjust classroom procedures to fit their own need and teaching style, as long as requirements set forth by law is preserved. Extreme care is advisable in removal of any policies included in the generic template as many of them are statements required by law. Their removal or alteration can cause exposure to serious institutional risk and is not recommended. Suggested additions to the standard policies outlined in the template revolve around establishment of a collaborative learning environment, on professional classroom behavior expectations and on specifics of instructor-student communications including modality and expected turn-around time.

The role of the course syllabi and a comprehensive course structure in overall classroom management effectiveness is best exemplified from the perspective of a business contract. Like contractual agreements between parties, the course syllabi needs to incorporate the terms and conditions related of the transaction at hand. The rights and responsibilities of both parties are to be clearly identified and defined and the expected actions and outcomes are to be meticulously explained to reduce the risk of misunderstandings. This leads to rather lengthy business contacts with extensive fine print clarifying the details of the exchange. The course syllabi also needs to be a comprehensive document that delineates the details of expectations, rights and responsibilities of both students and the instructor. It is, in essence, the agreed-upon standard for the quantity and quality of the educational outcome to be delivered and is used for solving any issues that arise from differences in perspective between the parties. Unfortunately, students often don't spend a lot of time reviewing the actual policies and procedures outlined in the course syllabi, focusing instead on the grading section. They glance through the policies to get a feel for what the course is going to be like, but they only check the details when they have specific questions or concerns.

A well built course syllabi can be rather lengthy, sometimes well over ten pages. Although not including details makes the document more approachable for students, the details serve both parties in case of misunderstandings or conflicts. In general it is advisable to treat the course syllabi as a living document adjusted from course to course, so that it can represent the specific course goals and structure the instructor wishes to establish and communicate. "The primary purpose of the course syllabus is to inform the students in a formal and timely manner of the nature and content of the course policies and procedures that apply. It reflects the instructor's professionalism and the quality of the course and describes the student learning outcomes, an overview of the course content, and the assessments used to evaluate the students' knowledge" (SCC, 2012).

The course syllabi starts with the course information section including the course title, delivery type, timeframe, meeting frequency and designated classrooms. This section is of foundational importance as effective instruction entails planning for the specific audience and delivery mode of the course at hand. While the other elements of classroom management relate to instruction in general, the course information section is dependent on and tailored to the exact details of the class. To proactively support those needing assistive technologies, a notice should be posted to contact the instructor or the institution's accommodations office directly if the need for special accommodations arises. This approach assists with self-advocacy which can be difficult for students who may fear being categorized as less capable or excluded from peer activities due to a condition.

Participants can be inexperienced or novice learners often seen in various computer, customer service and basic communication skills courses. Alternatively, the audience can be composed of highly specialized adults such as returning professionals taking continuing education or recertification credits or those changing careers in virtually any subject matter. Most often, however, the post-secondary career and technical education and training learners represent a healthy mix of diverse experience levels, ranging from age sixteen to fifty plus students. In comparison to K-12 and university education, where the student group tends to be somewhat cohesive, technical colleges and training providers attract a highly diverse population. This is due to the overall nature of instruction as well as their specific setting. In consequence, it is a best practice to proactively plan on the fact of learner diversity for classroom activities. So, if K-12 and university educators can design for a specific age-range with age-appropriate knowledge and experience, instructors and trainers should count on a wide range of ages and experience levels in their classrooms. Same goes for students' emotional maturity level, instructors having to design curriculum for diverse preparedness levels and attitudes. These aspects add yet a more complex dimension to the learning style preference all teachers face, making the job that much more challenging.

Like most aspects of diversity, the varied nature of career and technical education instruction is highly rewarding and generates personal and professional growth. However, managing global aspects of our environment is also time consuming and requires skill, will and a generous amount of patience. Thus, instructors should proactively understand the characteristics of their classroom audience and plan their course content and structure accordingly. When faced with diversity, teachers should maximize the benefits of having experienced, knowledgeable and mature professionals in their classrooms by using them as group mentors or project guides that can help less experienced peers. By incorporating emotional strategies such as peer mentoring or role modeling, instructors increase the engagement of all learners involved. While novice students benefit from the expertise and practical experience of their mentors, more advanced learners increase their skill by taking their knowledge to mastery levels only achievable by teaching others. A practical classroom example relates to continuing education credit professionals who when mentoring, in addition to reviewing course materials and refreshing their own knowledge, also rephrase and articulate the content to help others understand it better.

Effective course syllabi and calendar development also involves planning ahead for the audience size. In this regard, the K-12, the two year post-secondary and the university educational settings show some similarities along with some divergences. In this case the K-12 and the technical college classrooms are more alike, being smaller, more personable and interactive, while the university model being characterized by lecture hall setting and facilitation. Audience size affects the level and style of communication practices employed within each specific environment. In the case of small group discussions, interactions with peers and educator are likely and easy to accomplish. In the large group setting, on the other hand, content delivery quickly becomes one way communication with no room for ongoing formative feedback. Resulting is a considerable difference in planning activities and assessments for and delivering content to a small group of twenty versus two hundred students, career and technical education and corporate training having the advantage of the more intimate collaborative setting in a learning community.

The next section contains instructor information. It starts with clearly stating the instructor's full official name along with a phone and room number, and any other contact details. For part-time or adjunct educators without a designated office a best practice is to list the departmental office as the main contact. In some cases, off-site instructors may choose to list personal cell phone numbers to increase availability to students. However, due to privacy and confidentially regulation, most institutions strongly discourage, if not forbid the use of personal email addresses or phone numbers as these are not protected by law the same way as those associated with the institution itself. On the other hand, in the era of cloud computing and a highly mobile society, use of email and at least one alternative contact is a must. Institutions looking to better serve their learners and increase student satisfaction are starting to adjust by advising instructors to establish an active online presence via personal web pages and social media tools. Visibility and the availability for on-demand contact and immediate feedback are essential for student engagement and long term success. Some educators are reluctant to embrace social media tools for student communications considering them more personal networking tools than professional communication venues. But, in reality, social networking with peers and learners has shown to improve learning due to best practice and resource sharing as well as focused and relevant discussions. Just like with business venues where visibility is necessary for success, in today's mobile age students expect educators to be when and where they need them. In response, career and technical education and training institutions are using built-in communication features of course management systems to help automate student announcements and emails. Many take the next step by incorporating notification flags to stay connected with students. These strategies have proved successful being communication platforms preferred by the learner audience and responding to customer demands. The general visibility provided by being available to students virtually provides an effective means to increasing student engagement and academic success.

It is a best practice to expand the instructor information section to include an instructor welcome message as well. The role of the educator welcome is to communicate clear expectations while creating a warm, welcoming learning environment. It motivates using affective engagement factors. An energetic, positive atmosphere is one of the foundations of successfully completing a course or program. As the new student group is a highly diverse group of individuals rather than a cohesive group, the educator, whose role is to serve each person, must be cognizant of the diverse needs, goals, learning styles, backgrounds, and knowledge levels etc. students have and create an inclusive environment for all. Thus, it is important to incorporate informational details and strategic priorities as well as a personal tone to cater to as many personalities as possible. A recommended best practice is to start with a warm welcome message and then move to the educator's teaching philosophy and course goals. This approach allows establishment of instructor authority, presenting his or her professional achievements while also allowing introduction into course specifics. The overview of course outcomes and benefits in context of career applications, and the structural setup of the course are essential for a good start on student engagement. The firmer, business-like tone for communicating ground rules and expectations correlates with the policies and procedures found in the course syllabi, restating the terms and creating a sense of order and influence. It is a best practice to extend an offer to assist at the end in order to close on a positive note and to invite open dialogue on any lingering issues. This strategy is similar to the customer feedback surveys used in business which aim to solve problems in their inception rather than allowing them to become widespread concerns. For career and technical educators, the offer to help provides the opportunity to notice and correct any issues that could potentially surface with more than one participant. As a general rule, instructors can assume that concerns such as software versioning, online access incompatibility or the need for assistive learning technologies could exist in any class. Thus, when they invite student questions and concerns, they are likely serving multiple students. Additionally, being open to dialogue serves instructors well by creating the possibility of correcting problems they weren't aware of.

The next course syllabi section concerns course competencies and core abilities. Course competencies denote the mandated course outcomes regulated by the institution's accreditation agency. They are usually formatted as a descriptive summary paragraph followed by a bulleted list of specific skills and knowledge elements learners have when completing the course. The list contains straightforward, clear language, including action verbs that describe the means of demonstrating the skill or knowledge in question. Due to being highly regulated, this wording must match exactly the information used in course or program publications. It represents the official notice and warranty students receive prior to enrollment making the educational institution legally liable for providing students with the exact skill set or knowledge disclosed in preregistration media. Naturally, as this information is publicly marketed, the legislature considers it the reason for students choosing a specific course or institution rather than any other available alternative. In consequence, it holds the college and educator accountable for delivering on the product quality promised to customers. For example, the syllabi can state that "upon successful completion of the course, the student will have done the following: analyzed the poetry, drama, and fiction; identified historical and cultural influences on American literature and composed coherent essays about American literature using textual evidence" (SCC, 2012).

Core abilities, on the other hand, represent skills independent of course or program content or even industry and career field. They are transferable between careers and programs and are fundamental for success in any educational or professional setting. Examples of core abilities are effective communication, creative and critical thinking, teamwork, collaboration and responsible behavior. An example is the statement of "students who are successful will improve in the following general education areas: social and personal responsibility, communication, critical thinking and technical literacy" (SCC, 2012). Although most colleges consider them essential for any student's long term success, some place a heavier emphasis on communicating and enforcing them, making them optional in some institutions and required in others.

The course policies and procedures section of the course syllabi is essential for effectiveness. It is a best practice to borrow it from the generic, institutionally approved template designed by experts to assure its compliance with educational laws and regulations. This approach is recommended because, much like in business operations where the length of industry experience and professional education leads to reliable operating practices, college administrators are similarly in much better position of creating a properly worded risk reducing document than novice instructors. In fact, attempting to develop this section alone is counter-indicated by prudent and proven educational practice.

Fully understanding and embracing the course's grading policies is essential for learners. It is a best practice to incorporate general grading statements explaining the overall guidelines for various assessment methods. Statements such as "various formative and summative assessment techniques will be utilized during this course, including research papers, project drafts, journaling reflections and multiple choice exams to provide students with equal chance to succeed;

the mix of assessment methodologies is utilized to cater to diverse student needs and learning styles and to allow all learners to successfully complete the course if complying with the expectations outlined in the rubrics attached" can be used to communicate the strategy followed.

The ultimate goal of grading is the assessment of student knowledge and skills, serving various learning styles all at once. It should incorporate a clear definition for each assessment category and use a logical, straightforward point system. It should, in all cases, state the exact due dates for each assignment and should contain a brief explanation and benefit statement for the assessment strategy used. The number of points assigned to each assessment depends on the value and importance of each assignment in the context of the course as a whole. Finally, to add another layer of proactive, preventative classroom management, it is helpful to include the assignment rubric for specific grading details and expectations, clearly indicating what procedure will be followed for late submissions. For example, in case of a research paper assignment, the instructor can explain that the assessment connects theoretical comprehension with practical applications and results in deep, long-term knowledge. The rubric can also state that the assignment represents fifty points of thirty percent of the total points possible and that it needs to be submitted electronically by a specific date. It is a best practice to incorporate, in addition to descriptive paragraphs, a table visual identifying the final grade points or percentages rates also. The final grades should be consistent, as much as possible, college wide, to avoid confusing learners with inconsistent expectations for seemingly similar circumstances.

The last section of the syllabi is the course calendar, which is a table that lists the various learning modules incorporated in the course. Including the module titles, the assigned readings, the learning activities and due dates as well as their value makes the calendar a comprehensive but user-friendly overview of essential course elements. In fact, due to this, it is a best practice to include it in the new student welcome package as a stand-alone document. This approach allows students to use this convenient guide as a list of action items for their efforts. To maintain flexibility and reduce risk of educational law suits, it is recommended that a disclosure on the instructor's right to adjust or change the course content or structure at any time and without prior notice is included. This is a proactive protective measure against unforeseen events such as accidents, snow days or work or family related conflicts that could lead to alteration of content delivery. It is also imperative to state that students who have awareness of not being able to comply with one or more course expectations or deadlines should bring the conflict into the instructor's attention at the earliest possible date so that alternative arrangements can be made for satisfaction of all course requirements. These course syllabi statements are equivalent to the disclosure statements found on business contracts and need to be treated seriously.

The course calendar as a stand-alone section of the course syllabi contract serves the function of a user friendly success roadmap with positive effects on student achievement. If the course syllabi document serves its legally binding purpose, the course calendar engages students and increases learning outcomes. This is because in addition to listing each lesson plan module, reading assignment and assessment with their due dates and point values, the document also contains motivating language relating to the benefits of the skills and knowledge learned in the course. Also, should the educator, due to calculated or unforeseen events, miss sharing certain course content later included in assessment quizzes or final projects as requirements, the educational institution is protected from refund or grade adjustment requests on the grounds of certain content being delivered at a different time or date or being omitted in its entirety.

The course calendar is comparable to a production roadmap used in corporate management. The timeframe assigned for each production stage represents, like in industry, the plan for product design and delivery, allowing some flexibility for calculated delays. An example in this regard is the delivery schedule of raw materials necessary for production. Basic plant logistics as well as contingency planning requires that a steady raw material flow plan and a small back-up reserve exist should a delay prevent materials from arriving on time to the production site. A well designed production roadmap has a built-in back-up plan able to combat some of the delivery delays without negative effects on assembly. Only realistic plans that are proactive in calculating moderate risks are feasible and beneficial for effective business operations. In rigid plans unforeseen events can cause significant negative impact on operations and profitability. For example, it can outline a tentative course schedule itemized by modules or lesson plans, stating that module one concerns an introduction to financial systems, module two focuses on the distinction between federal and state regulatory entities, etc.

The course calendar can also be viewed as a visual of the product mix of overall course outcomes. In this perspective, the various assessments included in the course and listed as line items on the course calendar are components of a strategically developed product mix designed to satisfy all needs. Some of the products are relatively independent of each other, allowing for their use even when other pieces are not utilized, while others are highly interdependent, needing completion of one for use of another. A classroom example for this is the submission of individual photographs as stand-alone assignments followed by the creation of a final capstone project that incorporates all elements. However, in reality, even with the seemingly independent pieces there often is an inherent but hidden interdependence related to the development that occurs naturally along the course timeframe due to peer discussions, collaborative effort and instructor explanations. In the previous example, the basic understanding of color or contrast in photography or learning to use the features of the camera must be done first, before any individual assessments can be created and submitted. Also, many times knowledge gained via a research paper assignment will show in the next examination in form of additional insight or higher level knowledge incorporated is short answer or essay test questions. In a banking industry example, if the customer product package includes a checking and a savings account, a debit card, and online banking access, but the customer only opens a savings account, then the utility of the debit card product is null and the utility of the online transfer service is greatly reduced.

In business, we are introduced to competitive product pricing and product life cycles early on, often in introductory business coursework. We learn that prices represent demand and the value assigned by the target customer base to the products companies offer for sale. The value attached by customers is directly dependent on their perceptions of quality and benefits of the product in question. In education, grading corresponds to product pricing. The points assigned to a certain assignment represent the value attached to its benefits. For the student user, the benefit of completing an assignment correlates with the number of points allotted by the instructor. In consequence, if there is a high value correlated with an assessment the student will perceive it as meaningful and thus worth completing. The correlation between price represented by grade points and quality is direct, so the more points, the more value-added the assignment. Or, in other words, an assessment perceived important generates higher student engagement, leading to a direct correlation between grade points and student engagement. Students are more motivated to perform well on an assignment of relatively high impact on their overall course outcomes.

As far as competitive product pricing within the product package represented by the complete course with all its assessment elements, it is obvious that important products will be priced higher. In the business environment essentials have stable and often relatively high price while optional products are priced more flexibly for increased appeal. A similar phenomenon is seen in the classroom as well, leading to the instructional design best practice relating to carefully thinking through the entire course and creating grading percentage rates that correlate to the relative value of each element. Basic knowledge and skill assessment categories should be assigned larger grade point values while optional pieces should use lower values. Foundational evaluation tools must keep their relatively high value in order to increase student motivation for completion as they represent the essential course outcomes the institution is accountable for. In the business world, this prioritization is represented by the consumer with limited resources who will focus his or her purchasing power on base products, opting for or against optional elements based on capacity.

One could argue that students are to learn everything in a course. No doubt, using the complete product line available will likely have more benefits than only choosing selected essentials due to limited resources. However, given that limited resources are a fact of life, the reality is that there are few customers who purchase all components of a product package. Most pick the ones most beneficial to them personally and omit the optional. Similarly, no matter how good the comprehensive course design is, there are going to be students who choose to be selective in their efforts due to limited time, energy, motivation, etc. For example, in retail banking many will only open a checking account due to lack of funds for a savings account and will choose to do their banking in-person even though electronic banking tools are available to them. This means that in the classroom educators need to be cognizant of the reality of limited resources and price products in a way that resembles their relative value compared to each other. This can be done by weighted grades where weights assigned to each classroom assessment represents their importance and value or by simple grading where essentials receive large point values. The higher grades lead to increased student engagement and will reserved to the key course goals and outcomes while grade values with less motivating power can be used for optional course elements.

The course syllabi contract ends with the confirmation of understanding dated and signed by each student in hard copy or electronically. This voluntary agreement of the two parties represents the legally binding aspect of the course syllabi contract. By signing, the parties agree to the terms and conditions stated in the document and declare themselves accountable for their responsibilities and for awareness of their rights within this legally binding relationship. The importance of having a valid signature is underlined by the fact that business contracts are hard, if not impossible to enforce without a valid signature and date. Similarly the course syllabi, an educational contract, cannot be used as a default document for terms and conditions if one of the key elements of legal enforceability is missing. Thus, instructors as well as students need to be fully aware that the signed course syllabi document means obligations similar to any other transaction contracted in their personal or work lives. Going one step further, instructors are cautioned to be cognizant of the fact that the learning outcomes stated via the course competencies and core abilities as well as the overall course structure and procedures are contractual terms that hold them responsible for delivering content as stated in the document. Similarly, by signing the course syllabi, students agree to abide by the terms and conditions relating to the course expectations and procedures, including assignment submission deadlines and classroom behaviors. The document content, along with the confirmation is a protection in case of various misunderstandings or unforeseen events, including perception differences that could arise between the student and teacher. In short, a well written, comprehensive course syllabi is more than just a classroom tool. It is an education industry contract that has the same weight and consequence as any other business contract signed by the negotiating parties.

The understanding of the syllabi is the confirmation that students have read and acknowledged the expectations spelled out in the document. If we consider the course syllabi a contract its content correlating to the terms of an agreement the only significant difference becomes the deliberate approach of directing the student customer's attention intentionally to the essential details. The introductory statement stating that "I certify that I have read and understood the course goals and expectations" is the acknowledgement that the contractual parties are committed to the contractual terms. One may consider that the understanding of the syllabi represents only the student's acknowledgement of his or her rights and responsibilities. In reality no agreement exists without two parties who willingly engage in the contractual relationship. Thus, the signature line although seemingly only evidencing one party's signature, is in actuality a closure on the commitments of both parties. In education, one party is the teacher who authored the document and the other is the learner who accepted the terms outlined. Like with any retail service provider, the company, upon receipt of the customer's signature on the contractual agreement, commits to delivering the service in accordance with the terms outlined in the document. Thus, in effect, the customer's signature seals both parties to mandatory obligations. However signatures don't only represent responsibilities; they also give rights to those contracting. For students signing means rights to accommodations, to grievance procedure, to equitable treatment, etc. Moreover, the contractual relationship also gives students right to quality education delivered in accordance with the terms outlined, right to receive fair grading and right to institutional support services such as use of the library, counseling, advising, tutoring, etc. The course syllabi contract also gives rights to the instructor who, like a service provider, contracts expecting a safe, respectful and healthy work environment in which his or her expertise and judgment will be supported by the employer.

An optional, but very beneficial aspect of the understanding of the syllabi section that brings classroom management practices close to leadership is the commitment to excellence statement that regards the willingness of both parties to put forth their best effort for satisfying their mutual understanding. A statement like this accomplishes the very important goal of commitment of reaching for the best possible outcome. Affective in its nature and subjective, this element can turn the otherwise dry and businesslike course syllabi into a document of student and teacher engagement. As adults with prior experiences learners have engrained in them that contracts are to be respected and that excellence is rewarded. By committing the parties will be inclined to expect more of themselves and reach higher than otherwise. The commitment to excellence psychological factor benefits the development of a successful, long-term working relationship between the educator and the students. It acts as an assurance that the service rendered is of the highest quality and that the intention of the relationship is long-term. Research shows that in both business and education, such approach is beneficial leading to increased student commitment level, higher quality courses and better classroom interactions.

An alternative approach to the basic course syllabi is the comprehensive, all inclusive document which is not only a contract of terms and conditions but also a handout containing details such as rubrics for assessments. In this scenario, the document contains a short description of the key learning points and competencies to be demonstrated for each evaluation with the various levels of mastery and appropriate point values. The explanations cover the entire range of characteristics expected to be part of the demonstration of competency or skill and knowledge level. For example, research paper competency categories are number of references, grammar and spelling, mechanics, content flow, etc. with each having several possible quality levels. Thus, the number of sources utilized by the researcher represents varying levels of effort including adequate, below expected or above expected while a certain number of grammar or spelling errors correlate with different levels of competency. Instructors are not limited to utilizing the analytic rubric style described above. Flexibility is provided by using holistic rubrics instead which evaluate the overall content or outcome. In this case, the rows of the table contain a detailed description of the project as a unit, while the columns describe the levels of mastery demonstrated by the student. Thus, for a research paper, the description would state something like excellent, acceptable or poor submission, each including a number of elements like the number of sources cited, overall grammatical correctness, content flow and coherence, etc.

The comprehensive approach to course syllabi goes above and beyond the more generic format wide spread within the educational industry. In this case, the course syllabi functions more like a product manual found in business which, rather than only stating basic facts also incorporates the terms and conditions of the sale including technical and quality specifications and product a warranty. Although time consuming to create and not always feasible, such approach reserves some excellent benefits for the parties. Relating it back to the business world and looking at it from a contractual terms perspective, the more specifics are included in the agreement, the lesser the risk of loss and law suits regarding the satisfaction of contractual obligations. This is because legally the keys to contract enforceability are voluntary acceptance, signature and date, and a written format presentable in front of the court of law. The more clear the terms at the time of confirmation with signatures, the lesser the possibility for differences of perception in expected delivery standards. Thus, when students sign and date their confirmation of understanding of the course syllabi, they state that they reviewed and accepted the requirements as outlined, allowing the parties to be in agreement on what represents baseline requirement and what is above and beyond effort and what does not meet expected quality standards. If misunderstandings arise anyway, both the instructor and college leadership can easily reference the course syllabi to review the details of the contract governing the transaction and act in accordance with the agreed-upon terms and conditions.

A well written course syllabi leaves little room for guesswork and is an effective risk and liability reduction tool. From this perspective, it is recommended that the assignment rubrics are included in the comprehensive document, so that all aspects are clearly defined and agreed upon. For example, if class participation and expected professional behavior are graded as stand-alone items within the overall class structure, and discussion board conversations are also part of grading, it is a best practice to create specific evaluation rubrics for them just as for mid-term or final projects. This allows the comprehensive course syllabi to serve its role and purpose well, representing, from both instructor's and students' perspective, an insurance contract for the quality of the education delivered. In essence, the submissions students produce need to be of a certain quality level, meeting clear technical specifications. If these are not met, the product will be rejected and the student will not succeed. However, similar to implied guarantees regarding a product's inherent quality and functionality based on its intended use, simply because a rubric is not included in the course syllabi does not lift the provider's liability for reasonably expected usability. But, just like in business dealings, written warranties are more enforceable in front of courts of law, resulting in a recommendation for utilizing the comprehensive course syllabi approach.

Chapter 2 Summary Overview

This chapter analyzes basic career and technical education and training classroom management practices necessary for effective instruction such as ground rules, instructor authority and interventions. It examines basic career and technical education instructional design strategies and assessment approaches. The chapter concludes by examining the role of the course syllabi and the course calendar in achieving instructional outcomes.

Chapter 2 Key Terms

Assessment Strategies
Classroom Expectations
Classroom Management
Course Calendar
Course Competencies
Course Syllabi
Employee Orientation
Ground Rules
Interruptions and Disruptions
Success Roadmap

CHAPTER 3

Strategies for Leading Career and Technical Education and Training Instruction

The objectives of this chapter are to provide instructors a basic overview of:

- the role of leadership in accomplishing instructional goals
- strategies used to motivate and engage classroom participants

3.1. The Role of Leadership in Accomplishing Instructional Goals

Leaders excel by skillfully combining existing resources, encouragement and vision to achieve effective outcomes. Leadership, defined as "a process whereby an individual influences a group of individuals to achieve a common goal" (Northouse, 2007) is demonstrated in any social and workplace setting where people gather for learning, producing, or collaborating. Often, leaders emerge naturally as their personalities define their actions and decisions and set the stage for their position within the group. Other times leadership is the result of an acquired skill set and actions that build trust and motivation.

Career and technical educators take a leadership role in the classroom while in business, trainers and managers are responsible for leading staff, change and performance. Educational and industry leaders alike set high expectations and trust their followers to reach them at a high quality standard. Their actions and reactions are intertwined with their surrounding and teams, gaining employee input and feedback for decision making and creating buy-in for team actions taken. Sporadic evaluation of the company's work atmosphere, the level of employee loyalty and motivation is beneficial but not sufficient for successful operation. Rather, communication between leaders and followers needs to be ongoing. Continuous understanding of staff perceptions via weekly check-ins and coaching performance management discussions reduces the risk of employee disengagement and resentment of management decisions.

To showcase the essential nature of leadership for organizational success, the term learning organization was coiled to express that in the 21st century corporate success depends on the organization's ability to change and adapt to new demands, which, in turn, is dependent on effective leadership. Organizational learning needs to happen at every level including front line, middle management and strategic leadership. It starts with new employee and on-the-job training and continues with ongoing performance management and development. For new employees, long-term success starts with the new hire orientation because it acquaints them with the corporate culture and essential operational and service practices. Providing direction and boundaries at the same time, the orientation training sets expectations and ground rules and acts like the career and technical education course syllabi. On-the-job training, the next step in corporate learning, entails a close working relationship with a mentor until the new hire becomes a fully operational contributor to team efforts. The guiding and mentoring relationship allows both parties to grow and learn together and leads to outstanding results based on long-term, successful behaviors. Mutual respect develops for each other's knowledge and experience and the ongoing sharing becomes an effective venue for open communication and organizational learning. The critical questioning process ultimately leads operational improvements, management being open to new approaches that result in increased operational efficiency.

As far as the actual process, Kouzes and Posner (1987) provide a summary highly relevant for our purpose. They explain that the five keys to effective leadership are to challenge the process, inspire a shared vision, model the way, enable others to act and to encourage the heart. Challenging the process means to continually seek ways to increase the organization's effectiveness and to advocate continuous improvement. Inspiring a shared vision relates to understanding the company's strategic goals and vision and being able to convey it to followers in a way that is meaningful and relevant to them. Enabling others to act has to do with empowering employees to find ways to do things better by critical thinking and ability to act. Modeling the way entails setting a positive and encouraging example for employees to readily embrace and follow, gaining energy and hope that the expectations can be accomplished. Finally, encouraging the heart goes beyond the accomplishment of every day organizational tasks to share a common sense of purpose while providing extra motivation and support (Kouzes & Posner, 1987).

Corporate training sessions targeting employee groups who directly utilize the new skills in their roles closely resemble the career and technical education classroom. The audience is very diverse and includes cross functional teams and a mix of authority levels. Collaboration allows participants to gain new perspectives and learn aspects they wouldn't encounter otherwise. Much like corporate leaders, instructors need to masterfully assemble skills, knowledge and talent to produce an engaging and collaborative learning environment in which individuals team up for success. Educators must, like leaders, gauge the general classroom atmosphere and the level of learning. As a group of people with diverse needs, goals and entry skills, learners will need just as much team building as work groups do in the business setting. Work teams are formed to accomplishing a specific goal and their common purpose overarches the individual differences existent between the experts. Diversity is one of the few constant characteristics of teams and members maximize their individual talents and strengths to produce better quantity and quality together than on their own. Learners build collaborative knowledge based on individual backgrounds, insights, skills and talents achieving learning outcomes that none of the individuals could have reached alone. Discussions, resource sharing and peer feedback are essential elements of reaching masterful results within the collaborative learning community. This beneficial cooperation is one of the reasons for using formative assessments in post-secondary career and technical educational. Ongoing evaluation of learner knowledge and attitudes and creation of open communication via feedback are similar to continuous improvement efforts implemented in corporate coaching. In fact they are needed for instructors to play their essential classroom leadership role. To lead the way, educators showcase their expertise in value-added student interactions and encourage student effort, formative feedback providing the context for effective, hands-on leadership.

In contrast, summative assessment plays a considerably different role in leadership, being associated with the corporate performance review tool which assesses the overall quality of the employee's development in context of the company's general strategy. Over time, yearly performance assessments can reveal developmental trends aligned with the company's strategic planning and can be used for effective employee placement and promotions. Similarly, summative classroom evaluations are capstones of the ongoing formative assessment strategy and their outcomes, testing the learner's overall knowledge and pinpointing special areas of talent or skill. For example, if a learner produces a final photography portfolio with outstanding nature shots while the profile pictures are of average quality, the student can be advised to consider nature photography as a possible career path. Over the course of a program sequence, these observations will help clarify the career roadmap of the individual.

The actual strategies used in career and technical education and training imitate each other in their goals and applications. Indifferently of their exact setting, these strategies aim to motivate and engage participants in reaching their desired outcomes by creating collaborative and purposeful groups. For example, if the goal of corporate training is to train employees on effective sales practices, it is beneficial to have front-line retail and internal support staff representation in order to gain a good understanding on each others' objectives and on how the operation of departments is affected by certain actions. A typical banking example is the loan training in which the retail division will always seek to close a sale while the loan processing department has to balance the risk of loan default with loan portfolio growth goals. Having both sides present allows the groups to share perspectives and gain reciprocal insights, a goal that can be accomplished with the Cross Functional Teams strategy. Another example is the application of the Snowball Effect technique for gaining staff buy-in into new corporate strategies. In this case employees are asked to select the top three key terms relating to the new strategy and place it in the middle of a page or a circular visual. Next, each small group representing a distinct corporate division discusses and shares with the large group the top three benefits of the new strategic direction to their specific area. For example, if the new strategy is to expand sales into a new retail market, the benefits relate to higher sales and production, increased promotional opportunities, more customer service positions and ultimately, a more profitable company. When employees think through these benefits, putting words around the positive effects of a new direction on their job security, career advancement and income, they quickly buy into the leadership's decision and actively engage in supporting it. The benefits shared by each division are placed one by one on additional pieces of paper and wrapped around the original ball to create new layers on the snowball representing the snowball effect of great future possibilities induced by the change. In effect, the change turns from something unknown into a promising opportunity that allows accomplishment of personal and corporate goals at the same time.

In conclusion, it is clear that career and technical education instructional strategies apply very well to corporate training and can be effective, with appropriate adjustments, in both settings. Flexibility and a creative approach turn most instructional practices into valuable corporate training tools. In fact, the opposite also holds true. Strategies used in corporate training and leadership are effective career and technical education tools as well. Next, the analysis looks at how career and technical instruction and corporate training closely resemble each other in their overall business management and leadership aspects. The next step is to discuss various career and technical education and corporate training strategies that can be reciprocally beneficial in both settings. The instructional strategies are presented in a three layered approach listing specific action items necessary for their effective application. The corporate training techniques have a two layered structure rather than the three layers used for classroom strategies, because in the corporate training setting the majority of the sessions are one day events that do not allow sufficient time for using the exact same structure as classroom lessons utilize.

3.2. Strategies Used to Motivate and Engage Classroom Participants

The following section describes twenty six career and technical education instructional strategies that create student engagement and increase motivation. Each strategy is presented in form of a detailed overview, a list of benefits and three levels of action items to be implemented when the strategy is used for instruction. In most cases, the third and highest level of structure is optional, many of the strategies being applicable and beneficial with just two layers as well. The total number of twenty six has been purposefully chosen to correlate with the letters of the alphabet, exemplifying that educational leadership is part of the A, B, C of career and technical education instruction.

A) Alphabet Soup

Overview:

The Alphabet Soup strategy is based on an sentence which summarizes the course's core content or outcomes. The sentence is made up of five or six key terms, each assigned to a different small learner group. The actual words and the sentence itself is not shared with the students. Groups are asked to research possible course related terms or concepts that could be the term assigned to them based on the letters they receive from the instructor. When the small groups reconvene, students identify the key term and then the large group rebuilds the outcome sentence. Individuals reflect on how the course will benefit them personally.

Benefits:

- ✓ Motivates students to review and research course terms
- ✓ Creates a connection between course outcomes and personal goals
- ✓ Allows playful, collaborative learning

Implementation:

<u>Level One Action Items:</u>

- Prepare an outcome sentence of five or six words which summarizes the course's core content or outcome. Do not share this sentence with the students.
- Divide the large group into five or six small groups and assign one word per small group keeping the actual term confidential.

- Provide learners with the letters that make up the word assigned to their group continuing to keep the actual word a secret.
- Have learners research, on their own possible course related terms that use the letters they received.

Level Two Action Items:

- Ask students to identify within their small groups, the key term assigned to them based on the letters they researched.
- Once each group determined the correct word or term, have the large group reconvene and rebuild the summary outcome sentence originally created.

Level Three Action Items:

- Require that each small group discusses the summary sentence on their own, analyzing its meaning and significance.
- Ask that learners write up a one page reflection on what the course can teach them and how will they use and apply this in their future careers.

B) Bridging the Gap

Overview:

The Bridging the Gap strategy connects a theoretical concept or principle with its real-world applications. It reduces the theory application gap and creates connections between prior and new knowledge by listing course terms in one column and their practical uses in the other. The applications are then sorted by categories such as process or industry, feasibility or benefits and a concept map is built to show the term and its practical correlations (Angelo & Cross, 1993).

Benefits:

- ✓ Links theoretical knowledge with practical applications
- ✓ Connects past experiences with new learning
- ✓ Allows development of evaluation and synthesis skills
- ✓ Increases motivation and engagement as well as hands-on learning

Implementation:

Level One Action Items:

- Select specific terms or concepts relevant for the current course topic.
- Ask students to connect the terms with their applications by writing them on two separate columns of a table.
- Have learners synthesize their knowledge by explaining the real life use of the applications they listed.

Level Two Action Items:

- List all applications on a whiteboard and have students sort them into categories by process, industry, feasibility, benefits, etc.
- Build a concept map showing the key concept or term in the center and its applications as subsets connecting to the center allowing for creation of secondary or tertiary connections in addition to the main categories.

Level Three Action Items:

- Explain the term and categories of usage including their relationships to improve long term retention.
- Have students summarize what they learned in a synthesis paragraph homework assignment.

C) Cluster Connections

Overview:

The Cluster Connections strategy expresses the benefits of following a career pathway by choosing jobs within a certain cluster. It also shows that skills and knowledge gained in a course applies to many different jobs within a career cluster, showing the multilateral benefits and applicability of learning. Students focus on one specific career listed within a cluster, selecting a job that they are most interested in pursuing. Next, they review the assigned chapters from the textbook or course content, to understand the overall goals and objectives and then match their choice with the most applicable course materials. The strategy creates a clear connection between their individual career interest and academic success and helps focus student effort on the most essential content to accomplish personal career goals.

Benefits:

- ✓ Expresses the benefits of following a certain educational or career path
- ✓ Showcases skill and knowledge relevance, applicability and benefits
- ✓ Focuses student effort on essential and relevant content

Implementation:

Level One Action Items:

- ▪ Bring in a career cluster visual used in the state of instruction and show how the course or program fits into it.

- Discuss the big picture benefits of following a career pathway including theory and application relationships.

Level Two Action Items:

- Have students focus on one specific job listed within a career cluster; they should select the job that they are most interested in pursuing.
- Have students review all assigned chapters from the textbook or course content, to understand the overall goals and objectives.
- Ask them to match their job choice with the most applicable course materials to connect individual career interest and academic success.

Level Three Action Items:

- Group students by their career cluster preferences and have them discuss the relevant course content areas.
- Express how the learning applies and helps the development of their future success roadmap.

D) Cross Functional Teams

Overview:

The Cross Functional Teams strategy connects various student groups to discuss topics of common interest. The activity exposes students to a wider range of perspectives, new contexts and unusual applications of the course materials as learners can be from different programs, institutions or even continents. Topics of discussion can be selected by course brainstorming for added buy-in. The strategy works well when topics are intriguing for all participants and when the supporting technology functions correctly with voice, video and online chat capabilities (Barkley, 2010).

Benefits:

- ✓ Exposes students to new perspectives, applications and career options
- ✓ Increases cultural awareness and global outlook
- ✓ Helps overcome isolation in distance delivery and builds peer relationships
- ✓ Works well in online and blended delivery courses

Implementation:

Level One Action Items:

- ▪ Set up locations and groups from different programs, institutions or continents.
- ▪ Select a common topic to discuss, showing the relevance of the course content for other programs, careers, and geographical areas.

Level Two Action Items:

- Ask groups to summarize their perspective and reflection in a one or two page follow-up.
- Conclude with an individual synthesis evaluation of the personal significance of what was learned.
- Have individuals bring things together by building a puzzle image or collaborative essay including insights, correlations, and relevance across disciplines and cultures, etc.

Level Three Action Items:

- Create cross functional teams from the various learner groups connected via technology.
- Have students share their personal reflections and then compare and contrast their findings in an interactive discussion.

E) Dual Introductions

Overview:

The Dual Introductions strategy is well suited for a first class or introductory module where participants don't know each other. Students pair up with the person next to them and use a short profile sheet to guide their questioning around each other's name, profession and company, experience and hobbies. Afterwards learners introduce each other to the large group with the information captured during their conversation and then find a new pair based on having something in common. This strategy is beneficial for developing a collaborative learning community and for increasing student social engagement.

Benefits:

 ✓ Creates a collaborative learning community
 ✓ Increases student social engagement
 ✓ Improves presentation and communication skills

Implementation:

Level One Action Items:

 ▪ Ask students to pair up with the person next to them and provide a short profile sheet to guide their questioning around each other's name, profession, experience, etc.

Level Two Action Items:

 ▪ Create an opportunity for students to introduce each other to the large group, sharing the information they captured during the conversation with their assigned pair.

Level Three Action Items:

- Request that they find someone they feel they have something in common with and have them sit down for an additional discussion to deepen the relationship.

F) Engagement Tree

Overview:

The Engagement Tree strategy works with personal career and life goals placed on the branches of a tree visual. Students think about their goals relating to their future career and select three items that they most want to achieve. Students place their choices on the branches of a tree visual drawn on a whiteboard, adding related course elements or benefits as the class moves along. At the end of the course students pick them off the tree and discuss in the large group how these motivating course topics prompted them to focus better and learn more. As they hear each others' perspective, they build a social learning based collaborative environment in which they recognize their accomplishments and efforts leading to success.

Benefits:

- ✓ Creates interest and motivation to review many course materials
- ✓ Provides a fresh approach to creating awareness of course related benefits
- ✓ Helps students focus their attention on the most relevant materials

Implementation:

Level One Action Items:

- Start by having students think about their goals relating to their future career and have them select three items that they most want to achieve.

- Draw a large tree with branches on a whiteboard and have students place their top choices on their own branch.

Level Two Action Items:

- Have students revisit the tree at end of every class or module to add one or two new things they found interesting or engaging in that course section or class relating to their goals.
- Use shapes like stars or apples on the tree for writing in the new terms.

Level Three Action Items:

- Request that learners pick off the tree and discuss, in the large group, how these items helped them get motivated and learn more.
- Have students share their perspectives to build a social learning collaborative environment between students who have similar experiences or interests.

G) Matching Stop and Go

Overview:

The Matching Stop and Go strategy uses a matching game structure and fifteen or more course terms. Once students research and match definitions with each term, the cards are mixed and then sorted in three categories such as easy, medium and difficult. Students are asked to solve the first five easy matches. Students cannot move to the next level until they correctly solved the first category questions, representing a stop. Once they solve all the questions on this level, they can move on to the next level, representing a go, and receive more points. The same process repeats for the third level as well.

Benefits:

- ✓ Incorporates body-mind movement for learning
- ✓ Increases student motivation by points and individual research
- ✓ Creates healthy competition between peers

Implementation:

Level One Action Items:

- Create a matching game structure by selecting fifteen terms relating to a reading assignment or course section.
- Assign individual students to find the official definition for each term.
- Write the terms and definitions on individual cards and create a full set of cards for each student.

Level Two Action Items:

- Sort the cards in three categories such as easy, medium and difficult and give students the easiest set to solve individually.
- State that students cannot move to the next level until they correctly solved the easiest category, representing a stop.
- Request that students move to the next level only after they solved all questions on a level; once they do, they can move on to the next level, representing a go for additional points.

Level Three Action Items:

- Repeat the same process until all students solve all three difficulty categories.
- Determine and reward the winner who completed all question the fastest.

H) Motivation Maps

Overview:

The Motivation Map strategy revolves around the application of course related knowledge in real life setting. Learners are assigned a certain city, state or country to research. They bring back course topic related career opportunities, successful businesses, and renowned professionals relevant for their assigned geographical area. The instructor brings up a map containing these cities, states or countries and students mark their territory on the map and share how the course skill and knowledge outcomes apply and are represented in that area. The strategy provides a visual of just how far worldwide learning can take students, increasing effort and engagement.

Benefits:

✓ Increases awareness of global career opportunities
✓ Creates motivation by showing global content relevance

Implementation:

Level One Action Items:

- Discuss course topic, goals and major learning outcomes during the first class.
- Have students share, one by one, how they intend to apply and use the knowledge gained in this class in real life.

Level Two Action Items:

- Assign a city, state or country to each student and have them research related career opportunities, successful businesses, and renowned professionals relevant in that area.
- Ask them to write down two paragraphs on the results of their research including major industries, business ventures, celebrities, etc.

Level Three Action Items:

- Bring up a map containing each assigned city, state or county and ask students to mark their area and explain the results of their research investigation.
- Show how far learning can take learners globally, financially as well as professionally.

I) Personal Engagement Inventory

Overview:

The Personal Engagement Inventory strategy is a simple procedure for gaining insight into what students hope to achieve during a course. The technique allows assessment of the degree of fit between personal objectives and instructional outcomes and between the importance or difficulty of the competency sets to be reached. It works well for flexible topics where instructors can adjust content to increase personal relevance or when student objectives can be incorporated in the overall outcomes. The strategy increases engagement and clarifies the students' developmental roadmap (Angelo & Cross, 1993).

Benefits:

- ✓ Clarifies personal learning goals and developmental roadmap
- ✓ Increases curriculum and activity relevance and motivation
- ✓ Results in matching learner and instructional objectives

Implementation:

Level One Action Items:

- ▪ Ask students to reflect on their main interests in general such as nature, business, helping others, etc.
- ▪ Have them identify two or three areas of interest and write down related career interests such as professions, career fields or job titles.

- Require that they study the profession and look for strategies that lead to career success in the particular field.

Level Two Action Items:

- Give students a short outline of the goals of the program or course and ask them to compare their interests with the goals and outcomes of the program or course at hand.
- Have them rate the intensity of their interest on a scale of five.
- Ask them to evaluate what the results mean and how will they go about better maximizing their personal interests within the current course and or program.

Level Three Action Items:

- Pair students with the person next to them to discuss their findings and provide each other with insights and suggestions on ways to maximize their interests within the course.

J)   Personal Skill Log

Overview:

The Personal Skill Log strategy uses a log of course topics and skills that allows students to rate their interest and knowledge or skill level in each category. Having this awareness helps learners put extra effort on areas where their skills are most deficient or the topics of high personal relevance. Students can be grouped by interest area or experience level and can be assigned mentors, if needed. Visualizing the skills or knowledge they already possess is motivational because it shows students that they already have competence relating to their field or career choices. This technique works best with flexible curriculum where assessments can be selected based on personal relevance, competency or interest such as capstone projects, electives or community learning events (Angelo & Cross, 1993).

Benefits:

- ✓ Increases student confidence and study effort
- ✓ Makes curriculum and instruction relevant and effective
- ✓ Identifies the level of fit between course and learner objectives
- ✓ Builds personal relevance via flexibility

Implementation:

Level One Action Items:

- List ten skill or knowledge elements students have in their toolbox when they start the course.

- Ask that students rate their skill level or competency in each category.

Level Two Action Items:

- Divide students in small groups preferably by program major or year of study.
- Have them compare notes and discuss how these skills can be applied and used in real life.
- Discuss the areas they already know related to the course or career for motivation.

Level Three Action Items:

- Have students reflect on strengths and areas needing improvement.
- Help learners build a clear roadmap based on current realities and upcoming opportunities for development.

K) Personality Pair-Up

Overview:

The Personality Pair-Up strategy helps form highly effective teams for group projects. Students complete a short five question personality test online and instructors pair them for teamwork according to their personality traits. The students in each pair are asked to start working on their collaborative project. Given their similar personalities, their approach to project completion will be similar likely resulting in a seamless and rewarding partnership.

Benefits:

- ✓ Works well in online and blended delivery courses
- ✓ Increases awareness of own personality and task preferences
- ✓ Results in improved team collaboration for group projects

Implementation:

Level One Action Items:

- Ask students to complete a five question personality test online prior to class.

Level Two Action Items:

- Pair students based on their personality style such as direct, systematic, etc.
- Ask pairs to connect and start working on their team project.

<u>Level Three Action Items:</u>

- Have learners discuss their experiences with the group work and other team projects they were involved in.
- Pinpoint what went well and was rewarding versus what was difficult in working with an individual of a similar personality preference.

L)   Problem Rotation

Overview:

The Problem Rotation strategy develops critical thinking and problem solving skills via realistic industry scenarios. Students are grouped in small groups and receive a case study problem to solve. The scenarios rotate and each is solved by each small group. The last round is reserved for evaluation of the answers given. The activity concludes with the creation of a new best solution by synthesizing the results provided by the small groups during the rounds. The activity works best for complex problems that have more than one possible answer and is well suited for online delivery (Angelo & Cross, 1993).

Benefits:

- ✓ Fosters evaluation and critical thinking
- ✓ Improves student problem solving skills
- ✓ Increases content relevance by use of practical scenarios
- ✓ Works for large class sizes and online delivery

Implementation:

Level One Action Items:

- ▪ Develop multiple realistic scenarios of similar complexity and difficulty level.
- ▪ Create small student groups of four to six learners for a social learning environment.
- ▪ Rotate scenarios for multiple opportunities to practice critical thinking and problem solving.

- Have student groups find solutions individually to each case study problem.

Level Two Action Items:

- Ask students to review all solutions provided for the scenario originally assigned to their group and critically evaluate the pros and cons of each answer.
- Have learners select the best solution and summarize the reasons for selecting that option.

Level Three Action Items:

- Build a new best solution collaboratively with the large student group based on the best elements found in each of the answers.

M) Puzzle Piece Mail

Overview:

The Puzzle Piece Mail strategy focuses student effort on investigating course related topics. Students are structured in small groups and assigned a certain topic or module to research. The topics are pieces of an overarching theme with multiple perspectives. Groups present their topic in class, including an interactive discussion and knowledge sharing. Next, the large group draws conclusions on how all the pieces fit together and creates a puzzle using representative visuals for each topic presented and their correlations. Individuals establish deep level knowledge by writing a one or two page document about what they learned (Barkley, 2010).

Benefits:

- ✓ Increases student engagement by allowing individual research on assigned topics
- ✓ Improves presentation and communication skills
- ✓ Allows for knowledge and experience sharing within the learning group
- ✓ Provides a big picture visual of the overall course outcomes and benefits

Implementation:

Level One Action Items:

- ▪ Group students into small groups or pairs and assign each group a certain topic to investigate.
- ▪ Ask students to research and present their topic within a given deadline.

Level Two Action Items:

- Have small student groups present their topic in front of the class to peer teach their colleagues.
- Require that the audience observes and discusses the topics by interactive questioning, debate and knowledge sharing.

Level Three Action Items:

- Draw a large overall puzzle image using representative visuals for the topics presented.
- Show on the whiteboard how the topics correlate and interact, creating the big picture of the overall course content.
- Ask individuals to write a one or two page document of what they learned, summarizing the course outcomes and their career relevance and benefits.

N)   Questioning for Tests

Overview:

The Questioning for Tests strategy is based on an ongoing conversation between two students who take turns solving problems aloud and listening. Students relate their existing knowledge to the new situation and try to diagnose the cause of the problem. The listener follows the steps of the solution finding process verbalized by their pair attempting to understand the reasoning used and providing feedback on correctness and logic of the problem solving procedure (Angelo & Cross, 1993).

Benefits:

- ✓ Improves student analytical skills and verbalization of knowledge
- ✓ Fosters metacognitive awareness and helps diagnose problem solving errors
- ✓ Works great with distance delivery courses

Implementation:

Level One Action Items:

- Choose a topic or process that has multiple solutions or logical steps to follow.
- Assign the critical thinking scenarios or real life problems to solve to student groups.
- Direct pairs to solve the scenarios by taking turns asking critical thinking questions and answering them, leading step by step toward the solution via logical reasoning.

Level Two Action Items:

- Ask that pair partners use follow-up questions to clarify grey areas, further deepen the critical thinking process and get closer to the solution.
- Have partners peer teach each other and write down, after the solution is found, a few questions they considered essential for the critical thinking solution finding process.
- Direct students to turn in their questions at the end of class to serve as a pool for the final test.

Level Three Action Items:

- Collect the questions, evaluate them and provide large group feedback on how they compare in difficulty and content to the instructor's expectation.
- Create mock tests using questions from the student generated pool to practice for the final test.

O) Rainbow Rally

Overview:

The Rainbow Rally strategy puts a spin on course content representation by color coding all course related content such as videos, reading assignments, discussion topics, research topics, quizzes and lecture materials with colors borrowed from the rainbow. Small groups are assigned a certain color, together making up the full array of the rainbow. Small groups review their assigned materials such as all videos or all articles and select their top ten favorites. One by one, each group presents their choices and their benefits to learning or career success. Next, individuals are asked to select their top five choices out of all the options presented and write a research paper based on these resources to deepen knowledge in these areas of personal relevance and interest.

Benefits:

- ✓ Facilitates the review of all course related materials and selection of the top choices
- ✓ Deepens knowledge by research of personally relevant topics
- ✓ Improves analysis and presentation skills

Implementation:

Level One Action Items:

- ▪ Designate a certain rainbow color for each content type such as orange for videos, blue for reading materials, etc.

- Assign a certain color to each small group and have them review these course materials by a specific date.

Level Two Action Items:

- Require that small groups select their top ten favorite resources belonging to their assigned color based on their quality, effectiveness and relevance.
- Request that each small group presents their choices and explains the logic behind their selection.
- Ask that the small group presentations use the group's assigned color as background to let the colors of the rainbow shine.

Level Three Action Items:

- Ask that after the presentations individuals review their notes, selecting their favorite five resources.
- Direct students to write a research paper based on the materials they chose, deepening their knowledge and creating personal meaning in these topic areas deemed relevant and of interest.

P)   Reflective Summaries

Overview:

The Reflective Summaries strategy increases students' synthesis and creative thinking skills by creation of a representative paragraph that summarizes the content and answers its what, whom, when, where, how and why questions. It helps learners paraphrase the content of the course module or material assigned. The summary paragraph provides teachers feedback on areas that need clarification. It works great in any field for understanding concepts or theories and is well fitted for in-person, blended or online courses alike (Angelo & Cross, 1993).

Benefits:

- ✓ Improves students' ability to summarize and synthesize new information
- ✓ Helps students grasp complex concepts by deep analysis of each section
- ✓ Makes original information easier to recall and builds lasting knowledge

Implementation:

Level One Action Items:

- ▪ Use reasonable and manageable content amount for the weekly assignments; specific reading materials such as a chapter or article per week works well.
- ▪ Set clear expectations for completing the assignments by specifying deadlines as well as the quality of content required.

Level Two Action Items:

- Have students summarize each assigned content section into a one paragraph summary.
- Ask that learners place paragraphs into a summary document with five headings to organize materials into categories based on representative key words.

Level Three Action Items:

- Require students to complete a summary reflection document by writing a cohesive summary for each key word or heading listed on the comprehensive document created in the previous step.

Q) Relevant Rhymes

Overview:

The Relevant Rhymes strategy aims to create relevant and easy to recall rhymes for difficult course terms or concepts. Students research and then paraphrase course terms and definitions to make them easier to understand. Next, groups create relevant rhymes for each term using the term itself and the relevant keywords selected from the definitions. The rhymes represent the meaning of the terms or course at hand and help students memorize difficult course concepts. The strategy is well suited for in-person and alternative delivery modes.

Benefits:

- ✓ Makes difficult terms and definitions easier to remember
- ✓ Creates meaning by attaching a personal approach to dry definitions
- ✓ Increases class collaboration and interactivity
- ✓ Results in a learning tool handout suitable for test preparation

Implementation:

Level One Action Items:

- ▪ Divide the group in as many small groups as textbook chapters or modules included in the course content.
- ▪ Have each group focus on one chapter, selecting two or three relevant terms or keywords.

Level Two Action Items:

- Have the groups research and write down the definition and terms and then paraphrase them using own words.
- Require that learners identify two or three representative keywords from the definition.

Level Three Action Items:

- Ask that groups create relevant rhymes for each term using the term itself and the relevant keywords chosen from the definitions.
- Compile and post the rhymes created on the learning management system to make it accessible to the entire group as a learning tool handout.

R) Role Model Profiles

Overview:

In the Role Model Profile strategy students write a brief, focused profile of a role model individual from their career field whose values, skills or actions they admire. The strategy teaches students to identify, assess and explain values and behaviors that are important to them and allows association of course content with career related benefits. It shows that a set of varied values and opinions exist and creates awareness of diversity in the classroom. It works great in ethics or professional courses where decision making and choices are part of preparation for the real world of work. The outcomes are improved if a clear rubric exists describing the expectations on completeness and clarity of the profiles (Angelo & Cross, 1993).

Benefits:

- ✓ Requires students to identify and assess their own values
- ✓ Relates role model behavior to career success
- ✓ Directs analysis and reflection beyond surface findings

Implementation:

Level One Action Items:

- Reflect on values and expertise most important for a specific career and explain them to students.
- Have learners find someone they know and admire to serve as their role model; the role model should

preferably be a successful professional from their career field.

- Ask students to set interview time and date with the role model.
- Plan ahead by creating five questions students will ask and look for research resources for the overall context of role modeling.

Level Two Action Items:

- Alternatively, involve the large group in preparing five essential questions to ask by brainstorming possibilities and selecting the five best options.
- Ask that students interview the role model individually and take notes. After they reflect on the insights they learned, students summarize findings in a one page essay.

Level Three Action Items:

- Have individuals analyze how the outcomes affect their own life and career and identify ways they can improve to achieve excellence and become as successful as their role models.
- Ask that students commit to one or two action items and determine a time bound roadmap or action plan.
- Have learners discuss ways to accomplish their goals and hold themselves accountable to the roadmap and action items by building in accountability checkpoints.

S) Role Play Performances

Overview:

The Role Play Performance strategy involves the minds as well as the bodies of the learners as they create live scenes or model processes to show what they know. Students may model customer interactions, transactions, operations, procedures, etc. being interesting and engaging change of pace from traditional activities. It is recommended to allow flexibility for students to assign roles within their small groups to complete the assignment and plan sufficient time for role play performances and evaluations as they take considerable time. It is good to use an evaluation rubric and record the performance for peer evaluation and review (Barkley, 2010).

Benefits:

✓ Raises student interest and engagement based on considerable flexibility in completing the role play activity
✓ Increases collaboration between group members and creates awareness of each other's strengths
✓ Shows the application of the theory or procedure in real life setting

Implementation:

Level One Action Items:

▪ Provide an essential course related scenario or topic to use as role play theme.

- Assign small groups randomly or based on the experience level, program major or location of the students.
- Let students assign roles within small groups as to best fit their strengths and interests; role examples are researcher, actor, editor, narrator, etc.

Level Two Action Items:

- Ask students to visit a business or institution where interactions similar to the ones expected are part of daily activities.
- Expect group members to build a realistic role play scenario script together based on their observations or past experiences.
- Communicate a clear evaluation rubric for key performance standards expected such as strategies incorporated, length of performance, professionalism of handouts, audience interactions, peer evaluations, etc.
- Ask that a hard copy of the complete script including the names of presenters as well as the assignment goals and conclusions is submitted prior to the performance.

Level Three Action Items:

- Plan ahead and allow sufficient time for role play performances and evaluations.
- Have the small student groups role play the scenario by acting out the script they wrote, mimicking the real interactions they observed at the business they visited.
- Ask that peers watch and evaluate the performance based on the rubric provided.
- Provide specific targeted feedback to the groups on the competency exhibited and areas of improvement.

T) Route Graphs

Overview:

The Route Graphs strategy involves the creation of a clear course structure with specific modules, representative titles and scheduled timeframes. A discussion on the course goals and structure follows, explaining the logic of each and their successive order and progression. Based on the timeframes allotted for each activity or module, instructors create a graph that shows how the course content flows and moves along toward completion, discussing how much is learned and accomplished to provide motivation.

Benefits:

- ✓ Creates a clear visual for the course structure and progression
- ✓ Increases motivation by showing progress to completion

Implementation:

Level One Action Items:

- Create a clear course structure with specific modules, titles and timeframes.
- Communicate the structure to students via the course calendar and the syllabi.
- Discuss the modules with the learning group during the first class, explaining the reasoning behind their specific progression.

Level Two Action Items:

- Create a graph using content as the vertical axes and timeframe as the horizontal axes.
- Mark the course start as a point zero and mark new modules at the appropriate location.
- Connect the dots to show how the course is moving along and discuss how much is learned and accomplished to increase motivation.

Level Three Action Items:

- When all points are connected and show a line graph of the course progression, ask students to take turns marking with green the areas where they benefitted most from, connecting learning with its personal benefits.

U) Shop or Lab Learning

Overview:

The Shop or Lab Learning strategy allows learners to gain hands-on experience with realistic work environments or machines and benefit from direct application of skills. Students are guided by the facilitator toward a prescribed outcome but can form own predictions and determine the particular way to accomplishing a task. The more advanced skills learners have or more experienced they are, the more freedom can be given for individual discovery. Ultimately, students are allowed to formulate their problem statements and learning goals, instructors serving as distant advisors rather than facilitators. This activity can be considered a prework to a comprehensive role play scenario assignment where students need to demonstrate their skills and knowledge.

Benefits:

✓ Fosters direct application and hands-on experimentation
✓ Allows individuals to practice and discover
✓ Involves discreet and extroverted students

Implementation:

Level One Action Items:

- Determine content, delivery method and the activities appropriate for the audience.
- Predict areas of possible concern and plan troubleshooting resources and procedures.
- Assure that students complete the prework to understand goals and safety rules.

- Use experiential laboratories for sciences, shops for electro-mechanics and office settings for business.
- Divide the students in small groups and assign specific times to use the shop or lab setting.
- Ask that students take turns role playing and practicing the roles available in the context.

Level Two Action Items:

- Regroup students and have them discuss their experiences with the roles.
- Have them evaluate the pros and cons of working in a specific job and team setting.

Level Three Action Items:

- Ask individuals to reflect on their experience, explaining their favorite role and why that suits them well.

V) Snowball Effect

Overview:

The Snowball Effect strategy aims to provide a visual for the essential course terms and their connections or benefits. The starting key word is selected by the instructor or by the collaborative learning group is discussed and then written on a piece of white wrapping paper and crumbled up. A new word is selected per lesson plan or module. As the class progresses the group adds additional representative words for each topic or module. The words are written on new pieces of colored wrapping paper to build a colored snowball. Step by step, the class builds a colored ball which grows weekly with new concepts learned and creates a snowball effect of new knowledge.

Benefits:

✓ Synthesizes the overall course content and outcomes
✓ Teaches by play and collaborative discussion
✓ Provides a memorable visual of course outcomes
✓ Builds a collaborative learning community

Implementation:

Level One Action Items:

- Select an essential key word for the overall course content or the first module.
- Discuss the meaning of the word and the reasoning for its selection.
- Write the word on a white wrapping paper and crumble it up.

Level Two Action Items:

- Have group adds additional representative words for each topic or module.
- Wrap each in colored wrapping paper and create a new outer layer on the ball.

Level Three Action Items:

- At the end of the course unwrap the unwrap the snowball and discuss what was learned, revisiting the original term of the snowball as well as the layers wrapped around it over time.
- Clarify terms and examine the connection existent between, like a final course review.
- Use the revealed words to create a final summary sentence paraphrasing the course topic or lesson plan content.

W) Success Wheel

Overview:

The Success Wheel strategy revolves around identifying best learning practices. From an individual reflection level the discussion moves to a small group setting where is further narrowed and filtered. Six practices are selected per group, being rated good, great and best and placed in a triangle shape. The triangles ultimately form a success wheel of best learning practices students can utilize to maximize their time and study effort.

Benefits:

✓ Provides a visual of learning strategies that maximize learning time and study effort
✓ Creates a collaborative environment in which students learn from each other
✓ Results in a practical handout created by the students for the students

Implementation:

Level One Action Items:

- Have students create a list of their top ten best learning practices such as using the study zone, note taking, keywords selection, etc.

Level Two Action Items:

- Form five or six small groups and have them select their top six favorite strategies from the individual lists.

- Ask students to rate strategies as good, great and best; there should be one best strategy, two great strategies and three that are rated good.
- Have learners transfer their choices onto the levels of the triangle using the bottom area for the good category, the middle for the great category and the top section for the best strategy.

Level Three Action Items:

- Have the groups draw their triangles in front of the class, forming, collectively a wheel of success by creating a pentagon or hexagon with three layers.
- Ask groups to complete the layers of the wheel by filling in their preferred strategies; the end result is a wheel of student learning strategies that lead to academic success.
- Save the image and provide it as handouts for students for at home reference.

X)   Teaching Top Ten

Overview:

The Teaching Top Ten strategy starts out with a laundry list of high impact teaching practices created by students. Learners identify items such as clear course expectations, ongoing feedback, regular announcements, timely grading, etc. The instructor reviews the list and adjusts curriculum as much as possible to make it relevant and flexile. The role of this strategy is to have students realize and put into words the value of their education to their future and career success.

Benefits:

- ✓ Connects the value of education with career success
- ✓ Creates awareness of effective teaching practices on an individual and group level
- ✓ Leads to personalized and flexible curriculum

Implementation:

Level One Action Items:

- ▪ Set up a student brainstorm that results in a laundry list of high impact teaching practices that help them learn.
- ▪ Ask the group to narrow the list down to the best ten strategies as perceived by the current learner group.
- ▪ Review the list and adjust curriculum as much as possible to make it relevant and flexile.

Level Two Action Items:

- Ask individual learners to select three items from the list that are personally most meaningful and helpful to them to better learn and succeed.
- Request that learners share their selection with the large group.
- Take notes on ways to personalize curriculum for each individual.

Level Three Action Items:

- Expect that learners write a one page paper on ways they maximize their learning based on the ideas heard from others.
- Personalize course curriculum as much as possible to best fit individual needs and goals.

Y) Terminology Trading

Overview:

The Terminology Trading strategy helps students develop deep level long term knowledge related to essential course terms. Instructors can choose the terms themselves or allow the learner group to brainstorm them. Learners receive two terms and two blank index cards and are asked to write the correct definition on the blank cards. Next, the cards are mixed in the large group and small groups receive random cards. Each small group has four cards per student. Within their small teams, students try to match the terms with definitions. Some cards will match while some will not have pairs, so groups take turns negotiating for the correct match for the cards in their possession by describing the term or the definition in their own words to the other groups. If successful, the opposite group hands over the matching card; if not the new group takes over and tries to gain a pair for of their own cards. The goal is to collect as many correct matches as possible per small group. The group with the highest number of matching cards wins.

Benefit:

- ✓ Keeps students engaged and attentive to details
- ✓ Provides hands-on practice in a playful setting
- ✓ Increases focus and allows high level knowledge retention
- ✓ Improves retention of difficult concepts and terms

Implementation:

Level One Action Items:

- Select essential course terms or let students choose them from the overall course content.
- Have students find and write the definition of the terms on index cards to engage knowledge transfer via body-mind movement.

Level Two Action Items:

- Form small groups of four to six learners each and mix the cards.
- Give four cards per student and have them try to match terms with definitions.
- Ask groups to take turns negotiating for the correct match of the cards in the group's possession and that did not have matches within their small group.
- Have student groups hand over the matching card if negotiation was successful or allow the other group to take the turn to negotiate.

Level Three Action Items:

- Suggest that groups track which group could have the card they need so that they can effectively negotiate in the next round and collect as many cards as possible.
- Have students improve their explanations to be able to gain control of the card they want.
- Repeat until all cards found their matches and determine winner.

Z) Triple Entry Journals

Overview:

The Triple Entry Journal strategy allows students to note ideas, applications and perceptions related to a specific course or theory. The journal uses a table with four columns, one listing the theoretical concept, the second the definitions, while the third the related application. The fourth column is for noting reactions and perceptions relating to the topic, expressing their personal significance and meaning. The strategy is beneficial for evaluating student attitudes and interests relating to a career or procedure, helping to identify what students focus on and why, promoting self-reflective learning and personal value development (Angelo & Cross, 1993).

Benefits:

✓ Allows transfer of knowledge from theory to real life
✓ Records student reactions to theory, content or situation
✓ Encourages formation of personal meaning and evaluation of perceptions
✓ Promotes self-reflective learning and deep level knowledge development

Implementation:

Level One Action Items:

- Have students select two to five key terms or course topics and have them create critical thinking analysis questions based on the readings.

- Discuss with the large group the topics and any lingering questions, then hand out the journal table.
- Ask students to write the term in the first column and the definition in the second column.

Level Two Action Items:

- Use small groups to brainstorm all the possible applications for each of the terms and have the large group debrief and share results.
- Have students write down the list of possible applications in the third column.

Level Three Action Items:

- Require that individuals complete their journals by reflecting on how each theory or application made them feel and what they meant based on their own experiences and interest.
- Have students evaluate the personal fit between concepts and their career goals and turn in their triple entry journals at mid-term and course end for formative feedback and guidance discussions.

The remainder of the chapter focuses on ten corporate training and leadership strategies that create employee buy-in and increase teamwork and motivation. Each strategy is presented in form of a detailed overview, a list of benefits and two levels of action items to be followed when the strategy is implemented. In some cases, the second level of structure is optional, many of the techniques being applicable and beneficial even with just one layer. The total number of ten has been intentionally chosen to assimilate with the term top ten or most recommended strategies to use in business.

1) Brainstorm Blizzard

Overview:

The Brainstorm Blizzard strategy is for creating team member buy-in and new ideas for team action planning. Team members are seated in a circle and are given a topic such as increasing sales or better customer retention or improved productivity. Everyone takes turns in a rapid succession saying one word that represents a means to accomplish the stated goal. The trainer captures each idea and allows two or three rounds per topic. After all topics are finished, repeats are transferred to a new document to be used as team action plan items.

Benefits:

- ✓ Engages every team member in team action planning
- ✓ Leads to good buy-in and employee engagement
- ✓ Results is a high quality plan due to the benefits of the collaborative brain

Implementation:

Level One Action Items:

- Seat team in a circle and provide the team with topics of discussion.
- Have participants take turns stating terms considered beneficial for accomplishing the stated goals.

Level Two Action Items:

- Compile the list by retaining repeats deemed beneficial for multiple action items or topics.

- Discuss their benefits and alternative opportunities of implementation.
- Create a final document of team action items to be used as a future roadmap.

2) Customer Profiling

Overview:

The Customer Profiling strategy involves employees being assigned two or three customers to profile by using the company's internal systems, online research tools and their networks to gather data. At the session the participants are given a profile form to complete individually placing the results of their research into the standard profile format the company uses. Next, employees compare the profiles they created and discuss similarities and discrepancies. They teach each other by explaining where they found certain pieces of information and how they fit into the given categories. Next participants synthesize by collecting the information from the individual profiles, creating a master document that represents a comprehensive image regarding the given customer. The technique concludes by discussing everyday uses of customer profiles, including the benefits of knowing the customer base better and sharing useful information with team members for increased corporate success. The strategy can be used as follow-up activity to a customer service or sales training in a team setting or during professional development sessions by creating groups that work in the same geographical area or product line. Computer and internet access is recommended or the strategy should be flipped so everyone comes prepared with necessary information.

Benefits:

- ✓ Maximizes the benefits of the collaborative brain
- ✓ Allows hands-on practice with customer profiling forms and resource systems
- ✓ Leads to improved team collaboration and communication

Implementation:

Level One Action Items:

- Assign customers to participants ahead of time and plan logistics to have online and computer access for the training session.
- Communicate and enforce confidentiality and be prepared to troubleshoot.
- Ask participants to complete the standard profiles at the session sorting information into useful categories.

Level Two Action Items:

- Have participants compare profiles and lead a discussion on similarities and discrepancies.
- Request that participants synthesize the information in one master document showcasing the benefits of the collaborative brain.
- Discuss the benefits of collaboration and learning from each other in a team setting.

3) Eagle's Wings

Overview:

The Eagle's Wings strategy involves a picture or video of an American eagle flying high above mountain tops, where seemingly little possibility for prosperity exists. The eagle displays the special attentiveness of a hunting bird, flying over the area of possible prey. Participants are asked to write down ten or fifteen words relating to what this image means to them. Individuals are asked to share the top three aspects representing the results of their reflection, both the word and the key meaning they assigned to the word in the context of the corporate management or leadership. As they listen to each other, individuals gain perspective on what leadership roles entail with the realities of benefits and hardships relating to decision making for long-term corporate success and the pros and cons of managing teams and people. The strategy works best for training managers or future leaders participating in a leadership development trainee program.

Benefits:

- ✓ Creates awareness of the pros and cons of leadership roles
- ✓ Builds peer networking between managers or future leaders.

Implementation:

Level One Action Items:

- ▪ Provide an image or video recording of an American Eagle flying high above the mountains.

- Ask participants to write down ten or fifteen words that come to their mind when seeing the flying eagle and define their meaning.

Level Two Action Items:

- Request that participants share the top three aspects representing the results of their reflection, including the term and its meaning.
- Facilitate a discussion on what leadership roles entail so individuals gain perspective on the functions of leadership.

4) Flex Topics

Overview:

The Flex Topics strategy allows trainers or leaders to openly discuss areas that can be flexed to achieve maximum employee engagement and good work satisfaction. The strategy asks individuals to write down, anonymously, two things they would appreciate for flexing their work time and that would provide them with good work satisfaction. They do not have to disclose their name or choices; rather the notes are drawn out of a hat one by one and put on an easel sheet in categories. Repeat topics will be discussed first. A discussion on non-negotiable items can follow, changing the situation into an opportunity for team mentoring similar to classroom ground rules. Boundaries, possibilities regarding implementation as well as non-likely areas are discussed openly, creating a shared sense of purpose and collaboration. When middle managers return to their individual teams and repeat the strategy, they choose two or three items from the overall list to discuss with their teams as possibilities for flextime.

Benefits:

    ✓ Increases employee satisfaction and engagement
    ✓ Creates a sense of collaboration and ownership

Implementation:

Level One Action Items:

- Create a collaborative meeting and ask participants to write down two things they would appreciate for

flexing their work time and that would provide them with good work satisfaction.
- Sort the answers provided and count them to find the high priority items.
- Discuss the most often repeated topics first as far as possible ways of implementation.

Level Two Action Items:

- Call a branch level meeting to discuss two or three flex options that increase employee job satisfaction.
- Listen actively to team feedback and adjust as necessary the final decisions and means of implementation.

5) Focal Point

Overview:

The Focal Point Strategy helps individuals and teams focus on what's most important rather than getting lost in daily tasks without purpose or prioritization. Individuals are asked to create a list of all their daily duties and bring it to the session. Trainers compile the list in front of the group and then present a list of top ten team action plan priorities in comparison to the list created by compiled individual responses. The team works together to select by marking with three, two or one stars the activities that are most important in light of the ultimate goal represented by the accomplishment of their team action plan. Ultimately, the list of essential duties will be narrowed down to a top ten "to do" list that will help guide people's daily activities and prioritization decisions.

Benefits:

- ✓ Focuses employee effort on key team action items and tasks
- ✓ Creates an opportunity to reduce or eliminate redundancies and non-essentials
- ✓ Opens communication between team members and improves team effectiveness

Implementation:

Level One Action Items:

- Ask individuals to create a list of their daily duties and tasks and bring it to the session.

- Compile the lists and then show a list of top ten team action items.

Level Two Action Items:

- Facilitate the selection of the most important activities in light of the ultimate goal represented by the accomplishment of the team action plan.
- Narrow the list to a top ten to do list which can help guide the team's daily activities and efforts.

6)   Market Trending

Overview:

The Market Trading strategy is for creating a well rounded short to mid-term business plan which accounts for current and future market developmental trends. Individuals are given the topic of market trending and the specific geographical area or product line and are asked to come prepared with two or three essential trends they found or know about regarding the specific area or product. Examples are new product uses, complimentary or replacement products, new industry or area competitors, etc. Individuals take turns sharing insights sitting in a circle along with five or ten sentences of explanation. The ideas are captured in the order they are presented. After each round, a summary paragraph is produced based on the agreed upon trends related to the topic at hand. Next everyone takes another round and the same process of sharing and capturing ideas is repeated. Additional rounds may be used if necessary. Finally, a one page document is created to represent a synthesized executive summary that can be forwarded to executive leadership for consideration and review.

Benefits:

- ✓ Improves the outcome of team action planning by incorporating fresh ideas
- ✓ Increases mutual respect and creates trust between participants

Implementation:

Level One Action Items:

- Seat participants in a large circle and have them take turns sharing their insights.
- Allow only one insight at a time with a short explanation and capture ideas.

Level Two Action Items:

- Produce a summary paragraph of the trends or insights gained regarding the topic at hand.
- Direct participants to repeat the sharing process and note the ideas presented.
- Create a one page document to present to executive leadership.

7)   On the Job Mentoring

Overview:

The On-the-Job Mentoring strategy is used in collaboration with experts that serve as mentors to new hires or with teams who have additions and need to recognize the value of embracing and supporting their new members. Individuals are asked to write down five to ten things they appreciate when learning new things or faced with new situations. Next, participants are paired up to discuss the content of their lists, finding several similar and some divergent items. Once the large group reconvenes and pairs share things they had in common and how these relate to their mentoring roles and increase the engagement and learning of their mentees. The activity concludes by the large group compiling a comprehensive list of all the things pairs had in common, eliminating duplicates and creating a top ten mentoring techniques handout.

Benefits:

- ✓ Improves mentoring skills and expert knowledge sharing
- ✓ Increases collaboration between mentors by development of a peer networking community
- ✓ Leads to increased mentee engagement and learning

Implementation:

Level One Action Items:

- Ask individuals to write down five to ten things they appreciate when learning new things or faced with new situations.

- Pair up participants to discuss the content of their lists, finding several similar and some divergent items.

Level Two Action Items:

- Reconvene the large group and have pairs share the elements they had in common and their benefits.
- Request a compiled list of top ten essentials for good mentoring to use as future reference.

8) Profit Sharing

Overview:

The Profit Sharing strategy is suited for groups of eight or more people and involves four whiteboard and colored markers. Individuals are provided a table of behavioral preferences and circle the items that are most representative of them personally. The results show the employee's preferred personality style noted with A, B, C and D. Next employees group according to their assigned letters and each group is asked to create an invitation to a customer event using any resources they have in their possession. Outcomes will be different by personality style for the level of details provided, colors and designs used, etc. on the party invitation. Once completed, the different personality styles are revealed and discussed to show that each is a different way of accomplishing the same goal. The activity concludes with disclosing ways the team's diversity ultimately benefits the customer's diverse needs and wants and how working together shares the profits of each other's strength within the team setting.

Benefits:

 ✓ Increases awareness around individual personality style preferences
 ✓ Improves understanding between team members

Implementation:

Level One Action Items:

- Provide individuals with a table of behavioral preferences and ask them to circle the items that are most representative of them personally.
- Score the assessment and assign letters to each of the four groups.

Level Two Action Items:

- Ask each group to create an invite for customers.
- Let groups present their invites and share their logic and thinking process and then reveal their personality styles.
- Discuss the insights and conclude by stating that teams share the profits of each other's strengths within the team.

9) Teamwork Trivia

Overview:

The Teamwork Trivia strategy works with small groups or pairs. Participants are handed a sheet of paper with questions like "How many team members have children in this group?" or "How many is the number of total children in our team?" or "What are the names of the towns where our team members live?" regarding to their team. Group members answer these questions individually and then compare their answers within the group. The strategy usually generates lively discussion and leads to discoveries relating to what people on the team have in common. The strategy creates a fun networking opportunity and can be used in conjunction with a team breakfast of pot luck lunch (Barkley, 2010).

Benefits:

✓ Creates a fun networking opportunity inside the team
✓ Increases interpersonal communication
✓ Improves teamwork and collaboration

Implementation:

Level One Action Items:

- Group individuals in small groups or pairs and hand them a sheet with questions.
- Ask them to answer the questions listed in relation to their teams and peers individually.
- Direct participants to compare their answers within the small groups and to discuss their discoveries in more detail.

Level Two Action Items:

- Request that small groups compile their findings and the insights gained via this activity.
- Share the compilation in the large group setting so everyone gets to know each other better.
- Create opportunities for further discussion around things people have in common.

10) Twin Peaks

Overview:

The Twin Peaks strategy involves small work teams who belong to one department and focus on a project or a goal together. The goal of this strategy is to clarify the "What's in it for us?" at the time of new leadership decisions. The team is called into a relaxed meeting and the trainer presents the new project or goal to be accomplished, focusing on its positives and not on the challenges it may bring. Members are given a few minutes break to process the news individually, being asked not to speak to each other about the topic but to think of one or two benefits for them personally or to the entire team. The team reconvenes and participants share their examples of how the project or goal could benefit them or their team such as use of strengths or interests, a new opportunity for accomplishment, reaching or exceeding goals, etc. This creates a sense of ownership of the new goal or project, participants being motivated by being an essential part in accomplishing the team goal. The facilitator concludes by stating that the benefits are twofold, for company and individuals alike, creating twin peaks of new heights only achievable by working together.

Benefits:

- ✓ Engages everyone in the team actions and outcomes
- ✓ Provides experience and perspective sharing between participants
- ✓ Creates a positive work atmosphere and team spirit

Implementation:

Level One Action Items:

- Call team members into a relaxed meeting and present the project or goal to be accomplished.
- Focus on the positives it brings and ask members to identify one or two benefits for them personally.
- Give participants a short break for personal reflection and processing.

Level Two Action Items:

- Reconvene and ask that everyone shares their examples and look for team related benefits and connections.
- State that the benefits are twofold and new heights can be reached by working together to achieve more via collaborative expertise.

The examination of the selective corporate training and leadership strategies used for increasing employee motivation and engagement is not intended to be comprehensive; rather it serves the purpose of sampling the breadth and depth of techniques available to facilitators for moving from a procedural approach to visionary leadership. When incorporated in daily activities, the strategies presented render participants who contribute to the outcomes of the team and organization by creating a sense of belonging and ownership and increasing effort and output. Corporate leadership techniques are applicable in the educational setting just as career and technical instruction strategies work for training and development. In the classroom there are no managers and subordinates and so the intensity of hierarchical power is reduced. However, in business, the realistic means of making a difference in employee buy-in and work satisfaction is the utilization of engagement strategies such as those described earlier. The art of instruction and leadership are closely related, utilizing similar best practices for leading teams to success. As an example, the Teamwork Trivia practice can be easily adjusted for classroom use by having students play the game of finding out more about each other and then sharing their new insights with the rest of the class. Also, the Profit Sharing technique works just as well in the classroom as it does in the business setting because the discovery of the different personality styles helps students relate to each other and work together in class activities or on group projects.

Chapter 3 Summary Overview

This chapter examines various career and technical education and training strategies used to motivate and engage classroom participants. It discusses the role of leadership in accomplishing instructional goals and includes topics such as student and employee engagement, interests and goals, personal significance and career roadmap. The chapter serves as a basic overview of creating a collaborative learning community and maximizing the benefits of teamwork.

Chapter 3 Key Terms

Corporate Leadership
Employee Buy-In
Goals and Motivation
Instructional Strategies
Interests and Benefits
Leadership Roles
Personal Significance
Role Models
Student Engagement
Teamwork

REFERENCES

Andersen, A. (2000). *HR Director—The Arthur Andersen Guide to Human Capital.* New York, NY: Profile Pursuit, Inc.

Angelo, T., & Cross, K.P. (1993). *Classroom Assessment Techniques—A Handbook for College Teachers.* San Francisco, CA: Jossey-Bass.

Allen, I. E., & Seaman, J. (2007). *Online Nation: Five Years of Growth in Online Learning.* Needham: Sloan-C.

Association for Middle Level Education. (2012). *Formative and Summative Assessment Strategies.* Retrieved on September 3, 2012 from http://www.amle.org/

Bach, S., Haynes, P., & Lewis-Smith, J. (2007). *Online Learning and Teaching in Higher Education.* New York, NY: McGraw-Hill Publishing.

Bailey, C. (2004). *Enrollment, Attainment, and Occupational Outcome Patterns of Subbaccalaureate Business Students: A National Analysis.* Retrieved from http://nces.ed.gov

Barkley, E. F. (2010). *Student Engagement Techniques.* San Francisco, CA: Jossey-Bass.

Bolles, R. N. (2011). *What Color is Your Parachute?.* New York, NY: Random House Publishing, Inc.

Boone, EJ. (1985). *Developing Proper Adult Education.* Retrieved on November 2, 2012 from http://www.astd.org

Bradford, P., Cross, K. & Major, C. (2005). *Collaborative Learning Techniques.* San Francisco, CA: Jossey-Bass.

Burchell, M., Robin, J. (2011). *The Great Workplace.* San Francisco, CA: Jossey-Bass.

Clark, D. R. (2012). *The Art and Science of Leadership.* Retrieved on August 1, 2012 from http://nwlink.com

Darbyshire, P. (2005). *Instructional Technologies: Cognitive Aspects of Online Programs.* Hershey, PA: IRM Press.

Goodman, N. (2011). *The Evolution of Training and Development: Defining the Next Generation of Leaders.* Retrieved on November 2, 2012 from http://www.astd.org

Gootman, M. E. (2008).The Teacher's Guide to Discipline: Helping Students Learn Self-Control, Responsibility, and Respect. *American Journal of Educational Journal.* 6, p.36.

Gordon, H. (2008). *The History and Growth of Career and Technical Education in America.* Long Grove, IL: Waveland Press Inc.

Greive, D. (2000). *Advanced Teaching Strategies for Adjunct and Part-Time Faculty.* Ann Arbor, MI: Part-Time Press.

Greive, D., & Lesko, P. (2011). *A Handbook for Adjunct and Part Time Faculty and Teachers of Adults.* Ann Arbor, MI: Part-Time Press.

Kaplan, R. & Kaiser, R. (2012). *Stop Overdoing Your Strengths.* Retrieved on September 3, 2012 from http://hbr.org

Kaye, B., & Jordan-Evans, S. (1999). *Love'em or Lose'em—Getting Good People to Stay.* San Francisco, CA: Berrett Koehler Publishers, Inc.

Key, J. (1997). *Research Design in Occupational Education.* Retrieved June 26, 2011 from http://www.okstate.edu

Kouzes, J. M. & Posner, B. Z. (1987). *The Leadership Challenge.* San Francisco, CA: Jossey-Bass.

Lehman, K., & Chamberlin, L. (2009). *Making the Move to eLearning.* Lanham, MA: Rowman & Littlefield Education.

Lehman, R. M., & Conceiqao, S.C. (2010). *Creating a Sense of Presence in Online Teaching.* San Fransico, CA: Jossey-Bass.

Lorenzetti, J. (2005). Lessons Learned about Student Issues in Online Learning. *Distance Education Report.* 9 (6), p. 1-4.

Lynch, R. (2000). *New Directions for High School Career and Technical Education in the 21st Century.* Retrieved on November 27, 2012 from http://www.calpro-online.org

Mathis, R., & Jackson, J. (2002). *Human Resource Management—Essential Perspectives.* Cincinnati, OH: Thomson South-Western.

Merriam-Webster Online Dictionary. (2012). *Thesaurus.* Retrieved on November 10, 2012 from http://merriam-webster.com

Moskowitz, G., & Hayman, J. L. Jr. (1976). Success Strategies of Inner-City Teachers: A Year-Long Study. *Journal of Educational Research. 69,* p. 283-289.

National Center for Education Statistics. (2011). *Career and Technical Education in the United States.* Retrieved on August 5, 2012 from http://nces.ed.gov

National Research Center for Career and Technical Education. (2004). *Distance Learning in Postsecondary Career and Technical Education: A Comparison of Achievement in Online vs. On-Campus CTE Courses.* Retrieved on July 31, 2012 from www.nccte.org

National Research Center for Career and Technical Education. (2009). *Community College Access and Affordability for Occupational and Nontraditional Students.* Retrieved on July 31, 2012 from www.nccte.org

National Teaching and Learning Forum. (2012). *Classroom Assessment Definition.* Retrieved on September 3, 2012 from www.ntlf.com

Northeast Wisconsin Technical College. (2012). *Ways of Learning—Online Delivery Overview.* Retrieved on July 5, 2012 from www.nwtc.edu

Northouse, G. (2007). *Leadership Theory and Practice.* Thousand Oaks, CA: Sage Publications Inc.

O'Connor, B., Brunner, M., & Delaney, C. (2002). *Training for Organizations.* Cincinnati, OH: Thomson South-Western.

Organization for Economic Development and Cooperation. (2012). Glossary—Special Needs Education. Retrieve on September 3, 2012 from http://stats.oecd.org

Oxford Dictionary of Business English. (1998). *Glossary of Business Terms*. Oxford, OX: Oxford University Press.

Pearlman, R. (2011). *Personality Styles for Dummies*. Hoboken, NJ: Wiley Publishing, Inc.

Rath, T. (2007). *Strengths Finder 2.0*. New York, NY: Gallup Press.

Rath, T., & Conchie, B. (2008). *Strengths Based Leadership*. New York, NY: Gallup Press.

Reference for Business. (2012). *Encyclopedia of Business*. Retrieved on November 10, 2012 from http://www.referenceforbusiness.com

Rose, &Meyer. (2006). *A Practical Reader Guide in Universal Design for Learning*. Cambridge, MA: Harvard Education Press.

Rose, Meyer, & Hitchcock. (2005).*The Universally Designed Classroom*. Cambridge, MA: Harvard Education Press.

Rowe, W. G. (2007). *Cases in Leadership*. Thousand Oaks, CA: Sage Publications Inc.

Sandhills Community College. (2012). *Course Syllabus Guidelines and Template*. Retrieved on September 22, 2012 from www.sandhills.edu

Sandstrom, J., & Smith, L. (2008). *Legacy Leadership*. Dallas, Texas: Coachworks Press.

Scarcella, J. (2012). *Philosophy of Career and Technical Education*. Retrieved on November 15, 2012 from http://coe.csusb.edu

Scott, J., & Sarkees-Wircenski, M. (2008). *Overview of Career and Technical Education*. Orland Park, IL: American Technical Publishers, Inc.

Senge. P. M. (1990). *The Fifth Discipline—The Art and Practice of the Learning Organization*. New York, NY: Doubleday Publishing, Inc.

Silverberg, M. (2002). *National Assessment of Vocational Education*. Retrieved on December 9, 2012 from http://www2.ed.gov

Universal Design for Learning Center. (2012). *Universal Design for Learning Definition*. Retrieved on September 3, 2012 from http://www.udlcenter.org

Ward, K. (2012). *Personality Styles at Work*. New York, NY: McGraw-Hill Publishing.

Wikipedia. (2012). *Online Encyclopedia—Training and Development*. Retrieved on November 12, 2012 from http://www.wikipedia.org/

Wisconsin Technical College System. (2011). *Who Do We Serve*. Retrieved on August, 2012 from www.wtcsystem.edu

Wisconsin Technical College System. (2011). *Wisconsin Technical College System Reports*. Retrieved on September 20, 2012 from http://www.wtcsystem.edu/reports.htm

Wlodkowski, R. J., & Ginsberg, M. B. (2010). *Teaching Intensive and Accelerated Courses*. San Francisco, CA: Jossey-Bass.

Wolfgang, C. H., & Glickman, C. D. (1986). *Solving Disciple problems*. Retrieved on September 14, 2012 from http://www.amle.org

Woodward, C. (1980). *Manual Training in Education*. New York: Scribners and Welford.